MEDIA AND COMMUNICATIONS - TECHNOLOGIES, POLICIES AND CHALLENGES

FUTURE OF AUDIO AND VIDEO MARKETS AS PRESENTED TO CONGRESS

Media and Communications - Technologies, Policies and Challenges

Additional books in this series can be found on Nova's website under the Series tab.

Additional E-books in this series can be found on Nova's website under the E-book tab.

MEDIA AND COMMUNICATIONS - TECHNOLOGIES, POLICIES AND CHALLENGES

FUTURE OF AUDIO AND VIDEO MARKETS AS PRESENTED TO CONGRESS

CARTER POWELL LONG
AND
FISHER MCDONALD
EDITORS

nova publishers
New York

For permission to use material from this book please contact us:
Telephone 631-231-7269; Fax 631-231-8175
Web Site: http://www.novapublishers.com

Library of Congress Cataloging-in-Publication Data

ISBN: 978-1-62257-764-4

Published by Nova Science Publishers, Inc. † New York

CONTENTS

Preface		**vii**
Chapter 1	Background Memo to the Hearing on "The Future of Video" *Committee on Energy and Commerce*	**1**
Chapter 2	Testimony of Charlie Ergen, Chairman, DISH Network. Hearing on "The Future of Video"	**9**
Chapter 3	Testimony of Robert W. Johnson, CEO, Sky Angel U.S. LLC. Hearing on ''The Future of Video''	**13**
Chapter 4	Testimony of David Hyman, General Counsel, Netflix, Inc. Hearing on "The Future of Video"	**25**
Chapter 5	Statement of Jim Funk, Senior Vice President, Product Management, Roku, Inc. Hearing on "The Future of Video"	**29**
Chapter 6	Testimony of Gigi B. Sohn, President, Public Knowledge. Hearing on "The Future of Video"	**33**
Chapter 7	Statement of David Barrett, President and CEO, Hearst Television. Hearing on "The Future of Video"	**47**
Chapter 8	Statement of Michael P. O'Leary, Senior Executive Vice President, the Motion Picture Association of America. Hearing on "The Future of Video"	**65**

Chapter 9	Testimony of Michael K. Powell, President and CEO, National Cable & Telecommunications Association. Hearing on "The Future of Video"	**71**
Chapter 10	Background Memo to the Hearing on"The Future of Audio"	**81**
Chapter 11	Testimony of Ben Allison, Musician and Governor, New York Chapter of the National Academy of Recording Arts and Sciences. Hearing on "The Future of Audio"	**87**
Chapter 12	Testimony of David M. Israelite, President and CEO, National Music Publishers' Association. Hearing on ''The Future of Audio''	**93**
Chapter 13	Statement of Cary Sherman, Chairman and CEO, Recording Industry Association of America. Hearing on "The Future of Audio"	**97**
Chapter 14	Statement of Jeff Smulyan, Chairman, President and CEO, Emmis Communications Corporation. Hearing on "The Future of Audio"	**103**
Chapter 15	Statement of Steven Newberry, President and CEO, Commonwealth Broadcasting Corp. Hearing on "The Future of Audio"	**111**
Chapter 16	Statement of Tim Westergren, Chief Strategy Officer and Founder, Pandora. Hearing on "The Future of Audio"	**121**
Chapter 17	Testimony of Christopher Guttman-McCabe, Vice President Regulatory Affairs, CTIA. Hearing on "The Future of Audio"	**125**
Chapter 17	Testimony of Christopher Guttman-McCabe, Vice President Regulatory Affairs, CTIA. Hearing on "The Future of Audio"	**129**
Chapter 18	Testimony of Gary Shapiro, President and CEO, Consumer Electronics Association. Hearing on "The Future of Audio"	**133**
Index		**141**

PREFACE

This book examines the future of today's audio and video markets as presented in testimony to Congress. Testimony is provided by the chairman of the DISH network; Sky Angel; Netflix; Hearst Television; National Cable and Telecommunications Association; National Music Publishers Association; Recording Industry Association of America; and Pandora, to name a few.

Chapter 1 – Background Memo to the Hearing on "The Future of Video".

Chapter 2 – Testimony of Charlie Ergen, Chairman, DISH Network.

Chapter 3 – Testimony of Robert W. Johnson, CEO, Sky Angel U.S. LLC.

Chapter 4 – Testimony of David Hyman, General Counsel, Netflix, Inc.

Chapter 5 – Statement of Jim Funk, Senior Vice President, Product Management, Roku, Inc.

Chapter 6 – Testimony of Gigi B. Sohn, President, Public Knowledge.

Chapter 7 – Statement of David Barrett, President and CEO, Hearst Television.

Chapter 8 – Statement of Michael P. O'Leary, Senior Executive Vice President, the Motion Picture Association of America.

Chapter 9 – Testimony of Michael K. Powell, President and CEO, National Cable & Telecommunications Association.

Chapter 10 – Background Memo to the Hearing on "The Future of Audio".

Chapter 11 – Testimony of Ben Allison, Musician and Governor, New York Chapter of the National Academy of Recording Arts and Sciences.

Chapter 12 – Testimony of David M. Israelite, President and CEO, National Music Publishers' Association.

Chapter 13 – Statement of Cary Sherman, Chairman and CEO, Recording Industry Association of America.

Chapter 14 – Statement of Jeff Smulyan, Chairman, President and CEO, Emmis Communications Corporation.

Chapter 15 – Statement of Steven Newberry, President and CEO, Commonwealth Broadcasting Corp.

Chapter 16 – Statement of Tim Westergren, Chief Strategy Officer and Founder, Pandora.

Chapter 17 – Testimony of Christopher Guttman-McCabe, Vice President Regulatory Affairs, CTIA.

Chapter 18 – Testimony of Gary Shapiro, President and CEO, Consumer Electronics Association.

In: Future of Audio and Video Markets ... ISBN 978-1-62257-764-4
Editors: C. Powell Long, F. McDonald

Chapter 1

BACKGROUND MEMO TO THE HEARING ON "THE FUTURE OF VIDEO"*

Committee on Energy and Commerce

To: Members and Staff, Subcommittee on Communications and Technology
From: Majority Committee Staff
Re: Hearing on "The Future of Video"

The Subcommittee will hold a hearing Wed., June 27, 2012, at 10:00 a.m. in 2123 Rayburn House Office Building on "The Future of Video." One panel of witnesses will testify.

1. Charlie Ergen, Chairman, DISH Network
2. Robert W. Johnson, CEO, Sky Angel
3. David Hyman, General Counsel, Netflix
4. Jim Funk, Vice President, Product Management, Roku
5. Gigi B. Sohn, President & CEO, Public Knowledge
6. David Barrett, President and CEO, Hearst Television Inc.
7. Michael P. O'Leary, Senior Executive Vice President, Global Policy and External Affairs, Motion Picture Association of America

* This is an edited, reformatted and augmented version of the testimony given on June 27, 2012 before the House Committee on Energy and Commerce.

8. Michael Powell, President and CEO, National Cable & Telecommunications Association

The Communications Act is woefully out of step with the state of competition and technology in video distribution and programming. Cable operators had 98 percent of the pay-TV distribution market when Congress passed the 1992 Cable Act amendments and 53 percent of national program networks were at least partially owned by a cable operator. At the time of the 1996 Telecommunications Act amendments, cable still had an 89 percent share and 44 percent of national networks remained vertically integrated with a cable company. By contrast, as of 2006, the last year for which the FCC has released data despite its statutory duty to issue annual video reports, cable's market share had dropped to 68 percent and only 15 percent of national networks were vertically integrated with a cable operator. The two nationwide direct broadcast satellite providers had 29 percent of the pay-TV market and were the second and third largest providers, which they remain today. Over the same interval of years, the percentage of homes with televisions relying exclusively on over-the-air broadcasting dropped from 36 to 27 to 13 percent. Meanwhile, the number of national program networks has grown from 106 to 565 between 1994 and 2006.

The market has continued to evolve since 2006. The 2009 DTV transition tripled the channels broadcast stations can transmit and some broadcast stations are also rolling out mobile TV. Wireless carriers are streaming video. Programmers and pay-TV providers are putting their content on smartphones and tablets. Within the last ten years, YouTube, iTunes, Netflix, Amazon, Hulu, Roku, Sky Angel and others have begun providing video over the Internet. Verizon FiOS and AT&T U-Verse each provide TV service to 4 million households. Video providers have more sources of content, programmers have more sources of distribution, and viewers have more options than ever before. This hearing will examine where the video market and technology are headed, whether regulation of the old guard still makes sense, and whether regulation should be expanded to new technologies and services. What follows is a summary of some of the issues that may arise.

RETRANSMISSION CONSENT AND PROGRAM CARRIAGE

Retransmission consent and program carriage deals are best left arranged by the respective parties and their viewers, free from regulatory intervention.

These deals involve negotiations between broadcasters or non-broadcast programmers for access to pay-TV distribution on one side, and cable or satellite operators for access to programming on the other. Both sides in the negotiation hold valuable assets and both sides should have the right to withhold those assets if they do not believe a proposed deal amounts to a "good deal". Only if both sides face a genuine risk that programming might be dropped can a true negotiation take place. Whether it is programmers asking for access to "must-have" distribution or distributors asking for access to "must-have" content, calls for "standstills," government-forced arbitration, or other regulatory interference should be seen for what they are: attempts by parties to get a leg up in a negotiation. In some cases a distributor or programmer changes sides between support for and opposition to government involvement depending on the contour of a particular deal and the state of the negotiation. This changing perspective by parties on the necessity of regulatory involvement provides the best evidence that a robust market exists and is without need of regulatory interference.

The alternative is to ask regulators to weigh the relative value of particular carriage and particular programming, a risky proposition. At best this shifts money from one company's pocket to another's. At worst, it discourages investment in distribution and programming and diminishes diversity of content. Why would distributors experiment with carriage of unique content if they have a legal right to the most popular content of the day? Why would a programmer risk capital creating new content if there is little "shelf space" left because distributors have adversely possessed the hit shows of the moment or other programmers have muscled their way onto every cable and satellite platform? Regulatory intervention potentially puts the government in the middle of every negotiation and may embolden parties to make larger demands or to hold out longer than they should, to the detriment of viewers.

Service providers and programmers are closest to their audience and in a better position than the government to weigh the costs and benefits of carriage deals because both of these parties have something to lose: viewers. The viewers, meanwhile, have other providers and plenty of content to choose from should the parties make a bad decision. And with video available in almost every market from at least one cable operator, two satellite providers, and increasingly a phone company or the Internet, the same content is often accessible from at least one other source. It may even be available for free over the air or on a lawful website. The number of options viewers have and the financial interest the parties have in making a deal are perhaps why the vast majority of negotiations are consummated.

There are provisions of the Communications Act, however, that already put the government's finger on the negotiating scale to the detriment of pay-TV providers. The "must-carry" rules grant broadcasters a right to demand carriage on a cable or satellite system, although without compensation. As a result, broadcasters without popular content—who might otherwise need to pay cable and satellite operators for carriage—elect must-carry and rely exclusively on advertising to generate revenue. Meanwhile, broadcasters with popular programming still elect retransmission consent because they can collect money from cable and satellite operators for carriage. The "basic-tier" requirement entitles broadcasters carriage on the entry-level programming tier that cable and satellite providers offer subscribers. The "buy-through" requirement prohibits cable and satellite operators from selling subscribers other tiers, like for premium movie and sports channels, unless the subscribers have purchased the basic tier. The "program access" provisions allow the FCC to place restrictions on the rates, terms, and conditions a vertically integrated cable operator or programmer may place on a carriage deal, including restrictions on exclusive deals and whether vertically integrated cable operators and programmers may even refuse to enter into a deal. All these provisions restrict to varying degrees the ability of cable and satellite operators to freely negotiate carriage. They also limit shelf space that might otherwise be available for non-broadcast programming. If any provisions should be revisited, these should, especially in light of the increased competition and changes in technology that have arisen since their adoption.

Other provisions, such as the network non-duplication and syndicated exclusivity rules, also limit the ability of cable and satellite operators to freely negotiate, but should not be revisited unless done at the same time as compulsory copyright provisions. The network non-duplication and syndicated exclusivity rules prevent cable and satellite operators from importing network or syndicated programming from out-of-market broadcast stations if the local stations have negotiated geographic exclusivity agreements with the rights holders of the programming.

These provisions operate in tandem with the compulsory copyrights, which restrict the rights holders of broadcast content from prohibiting the content's use and set the royalties, if any, the rights holders may collect. The network non-duplication and syndicated exclusivity rules' restrictions on cable and satellite operators' ability to freely negotiate broadcast program carriage ameliorate the compulsory copyrights' restrictions on the rights holders' ability to freely negotiate use of and compensation for the content. That is why any re-examination should occur together.

SATELLITE TELEVISION REAUTHORIZATION

The retransmission of broadcast signals by satellite television providers and the copyright royalties they pay, if any, to the rights holders of the underlying content are governed by the Communications and Copyright acts. A number of the provisions periodically expire, requiring Congress to reauthorize them. The most recent reauthorization, enacted through the Satellite Television Extension and Localism Act of 2010 (STELA), expires Dec. 31, 2014. Among other things, the satellite television authorization in the Communications Act contains: 1) the "carry-one, carry-all" must-carry provisions requiring satellite providers to carry every local broadcast signal into the local market if it carries any; and 2) the distant signal provisions restricting satellite providers' retransmission of out-of-market broadcast signals into a local market to households that cannot receive local signals over the air.

"THE HOPPER"

FOX, NBC, CBS, and ABC are in a legal battle with DISH over whether the satellite provider's "Hopper" digital video recorder and PrimeTime Anytime service violate copyright law and their contracts with DISH. The Hopper, when used in conjunction with DISH's PrimeTime Anytime service, stores ABC, CBS, FOX, and NBC primetime programming so that subscribers can watch at any time any of the shows that have aired in the last eight days, as well as "auto hop" over the commercials starting the day after programs air. Subscribers can also save the programming indefinitely so long as they don't run out of room on the Hopper. DISH argues that the recording and ad-skipping features are permissible "fair uses" under copyright law. The networks argue that the storing, recording, and commercial skipping goes beyond the "time shifting" that the Supreme Court ruled was a fair use in the *Sony Betamax* case over VCR recording. They argue that the automatic skipping of commercials jeopardizes the financial underpinnings of broadcast television. The networks also argue that the service amounts to the provision of an unauthorized on-demand service in violation of the specific retransmission consent and video-on-demand agreements they have entered into with DISH. The networks filed a similar suit in 2001 against SONICblue over the ad-skipping feature of its ReplayTV digital video recorder. SONICblue went out of business before the court rendered a ruling.

Definition of Multichannel Video Programming Distributor

A company called Sky Angel has prompted a proceeding at the FCC that could result in the regulation of Internet-based video providers and content. Sky Angel provides subscribers a package of 80 program networks using a set-top box and subscribers' Internet connections. Many of the networks are the same as carried by cable and satellite providers. After getting into a carriage dispute with Discovery Networks, Sky Angel sought to file a program access complaint with the FCC. Because Sky Angel would need to be a traditional pay-TV provider—referred to in the law as a "multichannel video programming distributor"—to bring a program access complaint, the FCC opened a proceeding to determine whether Internet-based providers meet the definition of an MVPD.

The Communications Act defines an MVPD as "a person such as, but not limited to, a cable operator, a multichannel multipoint distribution service, a direct broadcast satellite service, or a television receive-only satellite program distributor, who makes available for purchase, by subscribers or customers, multiple channels of video programming." The Act defines a "channel" as "a portion of the electromagnetic frequency spectrum which is used in a cable system and which is capable of delivering a television channel (as television channel is defined by the Commission by regulation)." The Commission's regulations define a "television channel" as "a band of frequencies 6 MHz wide in the television broadcast band and designated either by number or by the extreme lower and upper frequencies." The Commission's regulations also define a "cable television channel" as a "signaling path provided by a cable television system." The Act defines "video programming" as "programming provided by, or generally considered comparable to programming provided by, a television broadcast station."

If the FCC concludes that Internet-based video providers make available for purchase multiple channels of video programming, they might fall within the MVPD provisions of the Communications Act. On the one hand, they might then receive certain benefits, such as the right to seek relief under the retransmission consent and program access rules. On the other, they might also be subject to certain obligations, such as the must-carry requirement, the basic tier and buy-through requirements, the requirement to enable the competitive availability of set-top boxes, the requirements to abide by the network non-duplication and syndicated exclusivity restrictions and negotiate in good faith with broadcasters for retransmission consent, the Equal Employment Opportunity requirements, the closed captioning and emergency information requirements, and various technical requirements. Consequently, this proceeding may have far-reaching consequences for the future of video.

In: Future of Audio and Video Markets ... ISBN 978-1-62257-764-4
Editors: C. Powell Long, F. McDonald

Chapter 2

TESTIMONY OF CHARLIE ERGEN, CHAIRMAN, DISH NETWORK. HEARING ON "THE FUTURE OF VIDEO"*

Chairman Walden, Ranking Member Eshoo, Members of the Subcommittee:

Thank you for the opportunity to testify about the future of video.

My name is Charlie Ergen and I am a Co-Founder and the Chairman of DISH Network, the third-largest pay-TV provider in the U.S. I also co-founded and am Chairman of EchoStar, a worldwide leader in satellite services and the design and manufacture of set-top-boxes. We serve approximately 14 million DISH subscribers and employ over 27,000 people throughout the U.S.

I believe that understanding the future of video goes hand in hand with two simple rules:

First, always try to understand what the customer wants.

Second, change is inevitable … embrace it.

What do customers want? They want to be able to watch programming on their TV sets, on their phones and on their tablets -- no matter where they are. They also want to be able to surf the web or make a phone call -- again, no matter where they are. DISH plans to offer consumers the chance to get all of these services from one company.

At DISH today, we do a good job of efficiently providing fixed video to the home. But customers increasingly want more than just home video. They want mobile video. They want mobile voice. They want mobile data. So, when

* This is an edited, reformatted and augmented version of the testimony given on June 27, 2012 before the House Committee on Energy and Commerce.

we look at the future of video, we need to be able to provide all of those communications services to every one of our customers, anywhere and anytime.

Our company is moving in that direction. With innovations like Sling, our customers can use a wireless smartphone or tablet from any location to enjoy the video content that they've already paid for. We recently purchased Blockbuster and are integrating their video content holdings to enhance our on-demand offerings. We are a leading distributor of DVR technology and continue to innovate so that our customers can watch the programming they've paid for whenever, wherever and however they want.

Our new "Prime Time Any Time" and "AutoHop" technology takes the DVR to a new level, giving consumers the choice to more easily view their preferred programming when they want, while skipping what they don't want to see. This means that allowing your kids to watch TV doesn't have to mean they have no choice but to see commercials for junk food and alcohol. Through AutoHop, DISH has done nothing more than improve upon existing, legally-accepted, and widely available technologies that give consumers the ability to record their television shows for playback at a more convenient time, when they are able to fast-forward through or skip over commercials.

These are some of the ways we have responded to our customers' changing needs. But we have further to go.

In the past, we haven't shrunk from "betting the company," so to speak, in order to stay competitive. We went from selling big dishes to launching our own small-dish, DBS business. To give customers what they want, including mobile video, voice, and data, we will have to take a significant risk once again.

Last year, we invested billions of dollars to acquire two bankrupt satellite companies, with the aim of transforming those assets into a next-generation mobile broadband service. We want to provide consumers with the choice in services and providers that they seek. If we're successful, we'll fuel billions of dollars in investment and create tens of thousands of new jobs throughout the United States.

And this brings me to my second rule: foster change; don't ignore it or be afraid of it.

We're prepared to leverage our experience and financial strength to drive communications and entertainment forward and make them more mobile and dynamic than ever before. We can't get started, however, until the FCC releases updated rules governing how our satellite licenses can be used for terrestrial mobile broadband. Given the overwhelming support of the comments received to date, we hope that the FCC will act, and finalize the new rules by the end of the summer. We want to build the most advanced wireless network in the U.S. to compete against the well-established incumbents, but we need to begin as soon as possible to have a chance.

Just as wireless rules must be modernized to keep up with consumers and technology, the rules governing how broadcasters and pay-TV providers reach retransmission agreements are outdated and in need of change.

Twenty years ago, when Congress first adopted the retransmission consent process in the 1992 Cable Act, there was typically only one cable operator in any given market, negotiating with only one broadcaster. Today, there are multiple pay-TV providers in each market, including satellite, telco, small local cable, and large regional cable providers. Then, of course, there are also the new over-the-top video providers like Netflix, Aereo, and others. The broadcaster still maintains a government sanctioned monopoly on network programming in his market, while pay-TV providers face stiff competition from one another. The result is almost always bad for consumers and the free market.

Broadcasters play the pay-TV providers against one another. They cut off the most popular sports and entertainment programming if their demands for drastically higher rate increases are not met. Consumers lose because they cannot see the programming they paid for, they end up paying higher rates, or both. And, the problem is only getting worse -- with more blackouts and more broadcaster abuses. From where we sit, the broadcasters cling to the status quo instead of meeting consumer demand and embracing new technologies and business models.

The retransmission consent regime is a prime example of an outdated government policy in need of an overhaul by Congress and the FCC. It is incredible to see how much has changed since 1992. Likely, there are few here who used the Internet back then, or had a cell phone. The idea of streaming movies or TV shows to a "smartphone" was science fiction.

Just as businesses must foster change in a rapidly evolving video marketplace to keep pace with what the consumer wants, government should work to ensure its regulations mirror today's competitive realities, consumer expectations, and advances in technology.

Thank you, and I look forward to answering your questions.

In: Future of Audio and Video Markets ... ISBN 978-1-62257-764-4
Editors: C. Powell Long, F. McDonald

Chapter 3

TESTIMONY OF ROBERT W. JOHNSON, CEO, SKY ANGEL U.S. LLC. HEARING ON "THE FUTURE OF VIDEO"*

Sky Angel provides an exclusively family-friendly, affordable service of about eighty linear channels of programming. The encrypted programming can only be accessed via a set-top box connected to a subscriber's broadband Internet connection. Because the set-top box connects directly to a TV set and subscribers select a desired channel through an on-screen guide, the service is functionally identical to traditional MVPDs. Like any new entrant, Sky Angel's ability to compete rests in large part on being able offer a variety of popular programming options. Recognizing this fact in 1992, Congress enacted the program access requirements, which prohibit a programmer affiliated with a cable operator from unreasonably refusing to deal with, or otherwise discriminating against, rival MVPDs. Unfortunately, the video distribution marketplace remains willing to engage in anti-competitive tactics in order to harm emerging competitors, as Sky Angel's ongoing experience clearly demonstrates.

Although Sky Angel had timely paid all fees to Discovery Communications for two years, Discovery suddenly informed Sky Angel that it planned to terminate their contract. Because Discovery refused to explain or retract its threat, Sky Angel filed a program access complaint with the FCC.

In response, Discovery argued that Sky Angel fails to qualify as an MVPD. Sky Angel fits squarely within the express terms of the MVPD

* This is an edited, reformatted and augmented version of the testimony given on June 27, 2012 before the House Committee on Energy and Commerce.

definition, is functionally identical to traditional MVPDs from a consumer perspective, and provides the exact competition Congress sought to encourage, but the FCC still has not ruled on the complaint twenty-seven months later, during which time Sky Angel has lost subscribers, and failed to gain new subscribers, due to the absence of Discovery programming. Anti-competitive conduct, and the FCC's failure to enforce the program access rules, continue to prevent the full realization of the public interest benefits Congress sought to achieve - increased competition, Internet use and consumer choice, affordable rates, innovation, and expanded programming diversity. Congress should act to ensure a level playing field for new entrants in the video distribution marketplace.

Chairman Walden, Ranking Member Eshoo, and members of the Subcommittee on Communications and Technology, it is my pleasure to have this opportunity to testify at your hearing today on The Future of Video and thank you for inviting me to participate. We think our experience with starting and operating an innovative new family-oriented video distribution service demonstrates significant problems with the manner in which the FCC is failing to enforce program access laws and regulations, so that a valuable new competitive entrant, such as Sky Angel, faces anticompetitive conduct from incumbent operators in the industry. We believe that the public interest in supporting competition in video distribution, expanded use of the Internet, diversity in programming sources, and affordable choices for American families requires the attention of the Congress.

INTRODUCTION: THE GOAL OF SKY ANGEL

Sky Angel was founded for the purpose of providing American families with a high-quality and affordable video distribution service in their homes that would offer family-friendly programming exclusively. I am one of the founders of Sky Angel. Our deep Christian roots guide the company. Sky Angel separately offers packages of Christian and secular video programming, as well as a combination package of all programming. Sky Angel is not a religious or proselytizing organization. Its goal is to offer a safe haven for families, and children, to enjoy as large a selection as possible of live video and audio programming, and recorded video programming, without fear of exposure to graphic sexual or similar content. Now operating under the brand FAVE TV, Sky Angel carefully selects all of its programming to ensure high-quality and strong family values. Our service is a proud recipient of the Seal of Approval from the Parents Television Council.

Starting in 2008, Sky Angel began distribution of exclusively family-friendly video programming, on a nationwide basis, at affordable subscriber rates. Today, it offers more than 80 channels that are live, linear programming networks, many of which are familiar in American households, such as The Hallmark Channel, Fox News, the NFL Channel, Bloomberg and the Weather Channel, to name a few. Sky Angel was the first American video programming distributor to provide a service which uses Internet Protocol Television ("IPTV") technology and a set-top box. Subscribers cannot access Sky Angel's encrypted programming without the set-top box, which has broadband Internet inputs and video outputs that connect directly to their television sets. No external computer or software is necessary to subscribe to, or receive, Sky Angel programming.

What Sky Angel Does and How It works

Sky Angel offers consumers a competitive alternative to other MVPD systems at affordable rates. Currently, the complete package of more than 80 live video and audio programming channels, as well as access to a large VOD library of exclusively family-friendly movies, costs $32.99 a month. American families can have Sky Angel in their homes as an exclusive source of wholesome video entertainment or they can subscribe and have Sky Angel in addition to another MVPD. The choice is theirs. All any American family needs for Sky Angel is a television set and a broadband Internet service of modest capability.

Although distribution of video programming through the use of IPTV technology is fairly well known in Europe, and other parts of the world, Sky Angel was the first to offer a true IPTV video programming distribution system throughout the United States using this technology. Sky Angel's IPTV system does not distribute programming on the World Wide Web, but rather relies on a combination of its own facilities, and in part, on subscribers' broadband Internet connections as a partial path in its distribution system. In summary, this is how Sky Angel provides its video service:

- Sky Angel enters into definitive written agreements with program providers/rights holders for distribution of programming to Sky Angel subscribers via its IPTV system under which Sky Angel pays the programmers monthly fees per Sky Angel subscriber who receives its channel.

- Sky Angel enters into subscription relationships with consumers for multiple, live linear channels of programming, who are sent Sky Angel set-top boxes, which are necessary to receive programming from Sky Angel and which Sky Angel directly and remotely controls at all times.
- Sky Angel receives content from programmers' satellite locations, and then processes and encrypts it.
- Sky Angel transmits the encrypted programming to its two headends via fiber optic cable which it controls.
- The encrypted programming then is distributed to Sky Angel subscribers, in part through Internet connections which those subscribers have contracted for from ISPs.
- The programming is received by the Sky Angel set-top boxes, decrypted, and then transmitted to subscribers' television sets with industry-standard copy protections.
- At no time is the World Wide Web, or home computers, part of the Sky Angel service.
- Sky Angel exclusively controls the origination, distribution, and reception of all programming, and at no time may anyone receive the programming except authorized subscribers via their authenticated set-top boxes.

To a consumer, Sky Angel is functionally identical to traditional cable and satellite video distribution services. All programming is available via a typical set-top box, which is controlled by the consumer using a standard remote control. The consumer simply scrolls across a menu of channel choices, and "clicks" the one he or she wants to watch. Sky Angel provides closed captioning, high definition and similar services on much the same basis as MVPDs do generally.

SKY ANGEL SHOULD BENEFIT FROM STATUTORY PROGRAM ACCESS PROTECTIONS

Congress enacted a broad definition of MVPD in the 1992 Cable Act, which the FCC implemented. That definition provides:

> ... the term "multichannel video programming distributor" means a person such as, **but not limited to,** a cable operator, a multichannel multipoint distribution service, a direct broadcast satellite service, or a television receive-only satellite program distributor, who makes available for purchase, by subscribers or customers, multiple channels of video programming... (47 U.S.C. §522(13) (Section 602)) (emphasis added).

Of course, in 1992 Congress did not envision video distribution partially using the Internet, because it did not exist, which is a perfect example of why Congress wisely did not limit the MVPD definition to existing 1992 technology. To qualify as an MVPD, and have statutory program access protections, an entity simply must offer multiple programming channels, for purchase, to subscribers or customers. By implication, there would be some similarity to traditional cable or satellite offerings, but this broad definition clearly is not specific to any particular technology or requires ownership or operation of specific kinds of facilities. Even a cursory look at how Sky Angel provides its service shows that it falls squarely under this definition of MVPD. And just as clearly, web-based, video-on-demand or non-subscription video offerings do not fall under the definition.

Sky Angel Faces Anti-Competitive Obstacles

Unfortunately, Sky Angel has encountered serious difficulties in the marketplace, primarily in its ability to acquire programming for its subscribers and to make the Sky Angel system more attractive and competitive. To be sure, Sky Angel does not attempt to attract viewers who insist upon access to salacious programming sources. But among the providers of must-have family-friendly programming, there is considerable resistance to entering into distribution contracts with Sky Angel.

Sky Angel adds programming channels to its system only through negotiated written agreements with programming owners. Under those programming distribution contracts, Sky Angel pays the programmers fees per subscriber which, in all likelihood, are considerably higher than the fees that large MVPDs pay, because a tiny start-up like Sky Angel has no market power or other ability to bargain for discounted fees.

A number of programmers simply will not deal with Sky Angel, essentially denying Sky Angel access to important programming sources, even though Sky Angel ends up paying premium fees.

The ongoing dispute between Sky Angel and Discovery Communications is a perfect example of the problem that Sky Angel faces in attempting to acquire, and pay for, must-have family programming. This dispute also shows that Congress needs to ensure that the regulatory processes it has put into place are being appropriately and fairly administered.

In 2007, Sky Angel and Discovery entered into an Affiliation Agreement for the distribution of five channels of Discovery programming, including "The Discovery Channel" and "Animal Planet."

Some Discovery channels are "must-have" programming for Sky Angel's service, and two of Discovery's channels quickly became among the top-five most-watched channels offered by Sky Angel. At all times, Sky Angel paid Discovery the monthly subscriber fees that Discovery wanted under the Affiliation Agreement.

The Affiliation Agreement, the term of which extends through December 31, 2014, expressly permits Sky Angel's use of IPTV technology, and provides a detailed definition of an IP System that is identical to Sky Angel's service.

In September 2009, Discovery proposed that the parties expand their relationship, asking Sky Angel to carry, and pay for, additional Discovery-owned networks. However, in December 2009, Discovery unexpectedly informed Sky Angel that it planned to terminate the Affiliation Agreement four years early, and thus withhold its programming from Sky Angel's subscribers and potential subscribers.

Although Sky Angel repeatedly sought additional information from Discovery, and offered to cooperate fully to address Discovery's alleged "concerns," Discovery refused to provide any justification, reasonable or otherwise, for its termination. Instead, Discovery simply repeated that it was "uncomfortable" with Sky Angel's "distribution methodology," which had not changed since Discovery carefully examined the system and the parties executed the Affiliation Agreement.

The definitions in the Affiliation Agreement provided in detail for the carriage of the programming over IPTV. Discovery was achieving its business goals under its Affiliation Agreement with Sky Angel - new distribution of its programming channels with payment of lucrative subscriber fees in return. Certainly, Discovery's sales personnel were pleased with the Sky Angel agreement because they sought to expand the arrangement, for additional fees, of course. However, for reasons that could not have been economic, Discovery terminated the agreement.

THE FCC PROGRAM ACCESS PROCESS IS AN ONGOING FAILURE

Because of Discovery's threats and refusal to explain or relent, on March 24, 2010, Sky Angel filed a Program Access Complaint and Emergency Petition for Temporary Standstill with the FCC in an attempt to prevent Discovery's unjustified and unlawful withholding.

The Complaint relied on the 1992 Cable Act and the FCC's implementing rules. On April 21, 2010, before Sky Angel timely responded to Discovery's claimed defenses, including that Sky Angel fails to "qualify" as an MVPD under the statute, the FCC's Media Bureau declined to issue a standstill, based only on the incomplete record before it.

Discovery severed its programming connection with Sky Angel the next day. The FCC has taken no other substantive action in the more than two years since April 2010, and so Sky Angel and its subscribers remain deprived of the Discovery programming or a decision on the merits by the FCC. Sky Angel has suffered harm, in the form of lost revenue, subscribers, and marketing opportunities, because of Discovery's action and the FCC's refusal to protect it.

Although the Bureau declined to grant a standstill at that time, it expressly did not rule on any of the merits of Sky Angel's program access complaint, including whether Sky Angel qualifies as an MVPD entitled to the program access protections Congress intended to apply to new, innovative competitors in the video distribution marketplace.

Rather, the Bureau simply found that, because of "the limited record before" it and "the lack of Commission precedent" regarding a service such as Sky Angel, it was "unable to conclude that Sky Angel has met its burden of demonstrating that the extraordinary relief of a standstill order is warranted."

The pleading cycle did not conclude until after the Bureau declined to grant a standstill, so Sky Angel had no opportunity to address Discovery's unexpected, and strained, legal and factual contentions made in opposition to Sky Angel's complaint and standstill petition.

The record of evidence before the FCC is that:

- Sky Angel qualifies as an MVPD entitled to relief under the program access rules. Sky Angel clearly falls within the broad statutory and regulatory definitions of an MVPD, and its innovative service, while not known to Congress twenty years ago, is exactly the type of new entrant Congress intended to promote in order to increase competition in the video distribution marketplace.

- Discovery continues to engage in unfair acts and unlawful discrimination by withholding its programming from Sky Angel while permitting far larger video programming distributors, including its affiliates, to offer their subscribers multiple linear channels of Discovery programming via broadband Internet connections.
- As has been demonstrated to the FCC, and is known to anyone who cares to look, DISH currently offers its 14 million subscribers unlimited Internet access to all of the live channels they subscribe to, including numerous Discovery channels, via Internet access from any U.S. location, at no extra cost. Discovery permits that Internet-based distribution of its channels by DISH. The "True TV Everywhere" offering by DISH proves that Discovery does not suffer harm from Internet distribution of its channels of a vastly greater scope than Sky Angel provides to its 20,000, or so, subscribers.
- Sky Angel has suffered, and will continue to suffer, irreparable harm due to Discovery's withholding of its programming. Sky Angel subscribers no longer can access Discovery programming, and consumers considering subscribing to Sky Angel have had less incentive to do so for more than two years. Further, Sky Angel's "first to market" advantage in offering this innovative and affordable service continues to deteriorate, and several large programming providers continue to hesitate to enter into distribution agreements with Sky Angel expressly because of Discovery's unlawful withholding.
- Discovery would suffer no harm because it would simply be providing programming to a distributor which uses the same distribution technology as other distributors of its programming, and Discovery would receive additional revenue on a per-subscriber basis at rates it required and contractually agreed to. Other large MVPDs also offer versions of "TV Everywhere" which include Discovery channels provided via some form of Internet distribution.

THE D.C. CIRCUIT HAD TO SCRUTINIZE THE FCC PROCESS

Under its program access procedures, the FCC has committed to resolving program access complaints within five months of submission. However, because of almost two years of inactivity by the FCC, Sky Angel felt compelled to ask the United States Court of Appeals for the D.C. Circuit to order the FCC to act in the case by filing a Petition for Writ of Mandamus. To Sky Angel's great satisfaction, in a matter of only a few days after filing its Petition, the Court ordered the FCC to respond to Sky Angel's Petition within 30 days.

Before the 30 days had expired, the FCC assigned the Sky Angel Complaint a file number, which the agency had failed to do for the previous two years. It also accepted a form of protective order between the parties, which also had been languishing at the FCC for most of two years. Then, the FCC "launched" a proceeding, asking for public comment regarding the meaning of the MVPD definition and asking many other questions, the vast majority of which were unnecessary to act upon Sky Angel's program access Complaint. The pleading cycle closed for those comments, and reply comments, on June 13, 2012.

To date, the FCC still has not acted substantively on Sky Angel's Complaint. Nor has it acted on Sky Angel's motion seeking information from Discovery (pending without action for more than two years), Sky Angel's renewed request for a standstill (pending without action for more than 13 months), or Sky Angel's request to sanction Discovery for lack of candor and to investigate if Discovery had misrepresented its claims of "harm" (also pending for more than 13 months).

CONGRESS SHOULD ENSURE THAT THERE IS A LEVEL PLAYING FIELD FOR VIDEO COMPETITION

So far, the regulatory process that Congress enacted in the 1992 Act has been an utter failure for Sky Angel, its subscribers, and the public interest which would benefit from the improved choice and pricing of its service. The fundamental purpose of Sky Angel's complaint is to evoke the program access rights under the 1992 Act so that Discovery would deal fairly with Sky Angel in its programming decisions, and not disadvantage Sky Angel, while providing exactly the same programming to large MVPDs. Sky Angel believes that the program access framework would be beneficial to it and the public interest if it were fairly and promptly enforced. Sadly, the small competitive entrant is being given short shrift by the FCC, and therefore, advancing the interests of multi-billion dollar incumbent MVPDs.

In fact, the recent record before the FCC demonstrates that a majority of interested parties, representing various segments of the video distribution marketplace, agree that Congress intended a broad definition of MVPD so that competitors, such as Sky Angel, would receive fair treatment in access to programming. In contrast, opponents are primarily incumbent MVPDs, who argue for a narrow definition of MVPD, taking exactly the opposite position they promoted in 1992 when the Act was passed and implemented.

Sky Angel, itself, is not a competitive threat to incumbent MVPDs because of its small size and family-friendly commitment. However, it represents a technology and system that could be very effective in competing against the established providers because of its far lower cost of entry into the market and ability to offer service nationwide, wherever broadband Internet connectivity is available.

In addition to the dispute with Discovery, Sky Angel has faced refusal to deal from other programmers. For example, and quite shocking, C-SPAN refuses to permit Sky Angel to distribute its channels. Although C-SPAN entered into an agreement with Sky Angel and began to provide its programming, C-SPAN cut its service off three days later, saying that it had "made a mistake," without any other explanation. Interestingly, C-SPAN is offered on the Internet as a "free" public service to all. Perhaps telling is that C-SPAN is controlled by the cable television industry, which claims that it does not need to be regulated in this area because it can be trusted to do the right thing. No doubt, the fox was quoted as saying much the same on the way to the chicken coop. In addition, a number of larger programmers, which are not distributors, have refused to deal with Sky Angel, saying (or implying) that they want to avoid conflicts within the industry.

Sky Angel believes that if Congress and the FCC made it clear that the law requires that it, and other new entrants, be treated fairly, then these failures to deal would be removed, or at least mitigated. Therefore, Sky Angel urges Congress to make sure that the existing provisions of the 1992 Cable Act are fairly and promptly enforced.

Sky Angel relies upon a set-top box as an essential component of its distribution system. Sky Angel supports fairness in set-top box technology and distribution. It would be delighted if set-top boxes generally supported the distribution of its service, so long as the integrity of its system, especially including encryption, is preserved.

Sky Angel is not an ISP. It does not provide Internet service. It relies upon the Internet connections of its subscribers for “last mile” access to the set-top boxes in their homes. Sky Angel has no arrangements with ISPs, and its service is subject to any data caps imposed upon subscribers by ISPs.

In conclusion, Sky Angel believes that it represents an important part of the future of video on the Internet. It offers a service that is cost-effective and consumer-friendly, permitting American families to have an alternative to the established, incumbent MVPDs, in a form that is functionally identical. It would improve use of, and access to, the Internet by consumers. The Sky Angel service is not web-based, or even connected to a home computer, and its characteristics fall clearly and squarely under the MVPD definition passed by Congress in the 1992 Cable Act. As such, on the one hand it should be protected under the program access rights that Congress included in the 1992 Act, and on the other hand, Sky Angel is entirely different from all web-based, video-on-demand and/or non-subscription video services, such as Netflix, Hulu, and YouTube. None of those services fall under the MVPD definition.

Sky Angel urges Congress to ensure that there is a level playing field so that it, and other new entrants, may have the opportunity to compete fairly for video subscribers and thus enhance the public interest, and consumer choice, by providing new opportunities and uses for the Internet, now and in the future.

Thank you again for the opportunity to testify and I would be pleased to answer any questions that you may have or provide additional information.

In: Future of Audio and Video Markets ... ISBN 978-1-62257-764-4
Editors: C. Powell Long, F. McDonald

Chapter 4

TESTIMONY OF DAVID HYMAN, GENERAL COUNSEL, NETFLIX, INC. HEARING ON "THE FUTURE OF VIDEO"*

Good morning Chairman Walden, Ranking Member Eshoo, and distinguished members of the Subcommittee. My name is David Hyman and I have served as the General Counsel of Netflix since 2002.

Netflix helped pioneer streaming movies and TV shows over the Internet. In 2008, we began to deliver instant streaming video to televisions through the use of a handful of Internet-connected devices. Today, over 23 million consumers in the United States use the Netflix streaming service on more than 900 different types of Internet-connected devices, including game consoles, mobile phones, and tablets. In fact, Netflix delivers close to a billion hours of streaming movies and TV shows to its consumers every month.

The Internet delivery of video provides consumers with unprecedented freedom and control over what video programming they can watch as well as when and where they can watch it. This is the future of video.

Increased demand for Internet video is driving consumer adoption of broadband. It is spurring a wave of innovation in the consumer electronics industry with virtually all new devices being Internet-connected or "smart." Innovation leaders, like Netflix, are developing new methods for delivering the petabytes of video data traversing the Internet. Just a few weeks ago, we announced the rollout of our Open Connect Network, a single-purpose content delivery network focused on the efficient distribution of large media files. Our

* This is an edited, reformatted and augmented version of the testimony given on June 27, 2012 before the House Committee on Energy and Commerce.

goal here is to help make a faster and less expensive Internet for all and to this end we are connecting to ISPs free of charge and making our hardware design and open source software components publicly available.

User interfaces, or UIs, through which consumers discover and engage with video are likewise evolving. The typical linear-channel selection grid that uses an up-down/left-right remote to access content will likely be replaced by intuitive and visually stimulating UIs that use motion, touch, and voice as means of control. Over the next few years, UIs will evolve in astounding ways. A decade from now, choosing a linear feed from a broadcast grid of 200 channels will feel as odd as would using a television dial to change channels today.

The Internet is helping to re-conceptualize how content is made and consumed. Last year, for example, Netflix announced it would be offering several original television series to its members. This is really the first time that such high-quality productions will debut through the Internet. One of our first shows will be *House of Cards*, a political thriller set right here in Washington DC. The series stars Kevin Spacey and Robin Wright and is executive produced and directed by David Fincher. We will be experimenting with how our programs are released, likely offering multiple episodes or perhaps even whole seasons at one time. In so doing, Netflix is giving artists like Mr. Fincher greater latitude to create compelling stories while giving our members the freedom to watch world class content in a way that suits their individual preferences.

Video services like Netflix are also leveraging the power of the Internet to help consumers discover content they will love. Using recommendation and merchandising technology, Netflix is helping to expand the reach and popularity of video content. For example, the availability on Netflix of all four prior seasons of the series *Mad Men* helped drive a 20 percent increase in those watching this season's opening episode. By combining technology with consumer choice, we were able to find an untapped audience of more than one million viewers for a show in its fifth season. This is just one example of how traditional and emerging video distribution platforms can complement one another.

All this change, however, has led to speculation about the demise of traditional video distribution platforms and networks. But it is our belief that these platforms and networks will also adapt to today's shifting video landscape. We see the beginnings of this with various authenticated Internet video offerings. Commonly referred to as "TV Everywhere," these offerings provide cable subscribers on-demand access to a variety of content through

Internet-connected devices like the iPad and Xbox. In this way, cable subscribers are afforded many of the benefits of Internet video within the bundled offering of their cable service. To get a good feel for this implementation, I would recommend you take a look at Comcast's Streampix, HBO GO, or Showtime's TV Everywhere offering. These implementations will likely grow in popularity and consumers will increasingly view traditional cable networks on devices connected to the Internet. At Netflix, especially as we focus on streaming and begin to offer original content, we are often asked if we are not becoming more like a traditional network such as HBO. The fact, however, is that these networks are becoming more like Netflix.

As traditional platforms and networks move to distribute their programming in an on-demand fashion over the Internet, they are beginning to compete more directly with pure-play or "over-the-top" Internet video providers like Netflix. As this trend continues, issues such as discriminatory data usage caps and IP interconnection must be examined with a much more discerning eye.

When you couple limited broadband competition with a strong desire to protect a legacy video distribution business, you have both the means and motivation to engage in anticompetitive behavior. Add to this mix a regulatory and legislative framework largely crafted before the modern Internet era and you have the makings for confusion and gamesmanship.

Competition leads to innovation and growing consumer choice. In large part because of innovation from Internet players like Netflix, traditional platforms and networks are changing their long-standing ways of doing business. In adapting to this changing landscape, these platforms and networks should not be permitted to unfairly leverage their data delivery networks or content distribution relationships to stifle unaffiliated video providers.

I encourage this Committee to examine closely the evolving competitive environment for Internet-delivered video. Netflix stands ready to work with you and others in the industry to explore various regulatory modifications or other changes that will help assure a competitive and innovative video marketplace for years to come.

Thank you for the opportunity to appear before you today, and I look forward to your questions.

In: Future of Audio and Video Markets ... ISBN 978-1-62257-764-4
Editors: C. Powell Long, F. McDonald

Chapter 5

STATEMENT OF JIM FUNK, SENIOR VICE PRESIDENT, PRODUCT MANAGEMENT, ROKU, INC. HEARING ON "THE FUTURE OF VIDEO"*

Good Morning Chairman Walden, Ranking Member Eshoo and members of the Subcommittee. Thank you for inviting me to testify today about the "Future of Video." My name is Jim Funk and I am the senior vice president of product management at Roku.

Roku was founded ten years ago in Silicon Valley. We are a rapidly growing company with less than 200 employees, still small by comparison with other companies represented at this hearing. The company's founder, Anthony Wood, is a serial entrepreneur who, among many accomplishments, started ReplayTV and is considered the inventor of the DVR. While the DVR is now among the most popular entertainment devices, Roku is betting on the Internet as the future of television.

Four years ago, Roku introduced a small Internet connected set- top box that allowed consumers to stream movies from Netflix over the Internet and watch them in high quality on their TVs without needing a computer. The combination of a $99 Roku player and Netflix streaming was a revolutionary offering. Now, four years later, there are hundreds of models of TVs, Blu- Ray players, game consoles, and set-top boxes offering this streaming video capability. These products can be found today at Best Buy, Wal-Mart, Target,

---

* This is an edited, reformatted and augmented version of the testimony given on June 27, 2012 before the House Committee on Energy and Commerce.

and Amazon.com marketed by well-known brands such as Apple, Microsoft, Sony, and Samsung.

Since the launch of the Roku player, we have sold more than three million boxes in the U.S. and now offer a range of very affordable products from $49 to $99. Roku has grown beyond a set-top box and has become an open streaming platform that allows content providers to create applications, which we call channels. Roku now features more than 500 channels that are available to the millions of households which have purchased our streaming players.

Most Roku users have a cable or satellite subscription service in addition to their Roku player, and Roku provides both new entertainment choices, as well as ways to get more value out of a cable or satellite television service.

In addition to Netflix, Roku users can now enjoy tens of thousands of TV episodes and movies on demand from services such as Hulu Plus, Amazon Instant Video, Crackle, HBO, and EPIX. We offer sports packages from Major League Baseball, the NBA, the NHL, and Major League Soccer, and there is news from NBC, Fox News, the Wall St. Journal and CNBC.

Beyond video, Roku users can enjoy streaming music services like Pandora, Internet radio from around the world, photo sharing, and popular games like Angry Birds.

Roku is becoming an alternative way to reach consumers for existing video distributors too. We recently announced a partnership with Dish Network to stream their international programming to consumers who may not be able to install a satellite dish. Customers of Time Warner Cable, Verizon, Dish, and other MVPDs can now enjoy on- demand movies and shows from HBO and EPIX on Roku.

Local broadcasters have even begun to experiment with reaching their audiences on Roku. Today you can watch local news on demand from broadcast stations located in Madison, Wisconsin, Las Vegas, and Indianapolis.

Roku is also a means for content producers who do not have traditional cable or satellite distribution to reach the living room TV via the Internet. For example, we have over 75 faith based channels, representing everything from individual congregations to Catholic and Mormon Church broadcasts. We even have a channel created by the House Oversight and Government Reform committee where video of hearings like this are available on demand.

We believe that devices like Roku are part of the future of television because of what we already see. The average Roku user streams over 10 hours of video per week, which is almost one-third the number of hours that the average American watches traditional TV. Consumers like the new technology of Internet streaming because it combines all the new content choices with convenience and value. With the widespread adoption of DVRs, consumers now expect on- demand viewing, and the Internet provides virtually unlimited capacity for delivery of on-demand content. In addition the combination of inexpensive Internet connected devices and the expanding selection of Internet video services offer an excellent entertainment value for cost- conscious consumers.

I did not come here today to advocate for specific legislation. Our point of view is that devices such as Roku represent an area of exciting innovation in entertainment and information delivery that are finding a home in the living rooms of millions of consumers. These devices are being embraced by all segments of the entertainment industry as a means to expand the business opportunities for legitimate content distribution. They are also driving the adoption of high speed broadband connections, to the benefit of ISPs.

Our interest here and we believe this is also the consumer's interest, is that there continue to be an open marketplace for competition in this space. That includes not only open competition between device manufacturers, but also open competition between video services, both traditional and new, as well as competition between Internet service providers. The widespread availability of affordable high speed Internet, open to all video sources is essential to continued growth and innovation in this market.

Thank you for your time today. I look forward to your questions.

In: Future of Audio and Video Markets ... ISBN 978-1-62257-764-4
Editors: C. Powell Long, F. McDonald

Chapter 6

TESTIMONY OF GIGI B. SOHN, PRESIDENT, PUBLIC KNOWLEDGE. HEARING ON "THE FUTURE OF VIDEO"*

Chairman Walden, Ranking Member Eshoo, members of the Subcommittee, thank you for inviting me to talk about the future of video. My name is Gigi Sohn, and I am President of Public Knowledge, a non-profit public interest organization that seeks to ensure that the public benefits from a communications system that is open, competitive and affordable.

INTRODUCTION

There is widespread agreement that we are currently living in the golden age of television. Programs like *Mad Men*, *Breaking Bad*, *Game of Thrones*, *Modern Family*, *The Daily Show* and *The Colbert Report* have become part of our cultural landscape, and even in this era of 500 channels, still inspire discussions around the water cooler. There are numerous new ways to watch TV – be it on a flat screen LED television, on a tablet or on a smartphone. And the Internet and DVRs have the ability to allow a viewer to watch what they want to watch when they want to watch it.

Despite all of the great programming and groundbreaking devices, Americans are locked into a television business model that limits competition

* This is an edited, reformatted and augmented version of the testimony given on June 27, 2012 before the House Committee on Energy and Commerce.

and choice; keeps prices for video high and limits technology and online video from achieving their full potential. This business model is made possible largely by an outdated regulatory structure created by incumbents to gain competitive advantage. It is time for Congress and the FCC to revamp this regulatory structure so that new video competition can thrive – giving consumers greater options and the ability to watch video whenever they want and on the device of their choosing.

Congress and the FCC can achieve this goal in three ways. First, they can clear away some of the outdated rules that slow down the evolution of the video marketplace. Second, they can extend the successful policies that allow competitors to access high-value content to certain online providers. Third, they can protect Internet openness and prevent discriminatory billing practices that hold back online video. By doing this they will increase competition, which will mean lower prices, better services, and more flexibility and control for consumers.

Background

For nearly a century the federal government has shaped the development of electronic media. In the 1920s the Federal Radio Commission brought order to the chaotic and experimental landscape that characterized early broadcasting. In doing so it set the conditions that allowed radio and then television broadcasting to develop into what it was in its heyday, and what it is today. In the 1960s and 1970s the FCC took steps to protect broadcasting from the disorganized and innovative early cable industry.[1] By doing this it made sure that cable became an adjunct to rather than a replacement for established broadcasting.[2]

After Congress passed the Cable Act of 1984, the tables turned and cable became the monopoly. Cable operators controlled who did and didn't get on the new medium, using their power to require cable programmers, such as the fledgling CNN and Discovery, to provide "pay for play" equity interests to cable operators, or sign exclusive agreements prohibiting programmers like MTV from appearing on potential competitors such as Direct Broadcast Satellite (DBS). At the same time, cable operators received access to needed inputs such as poll attachment rights and broadcast programming. The lack of competition led to high prices and poor service, but the cable incumbents' control over "must have" programming made it impossible for any competing services to emerge. It was not until the 1992 Cable Act[3] that Congress embarked on an express policy of increasing competition in the television market. It realized that potential competitors needed access to the same content as large cable systems with market power. New laws such as program access rules that gave competitors access to programming owned by the cable operators, and program carriage rules that prevented cable operators from demanding an equity share as a condition of carriage ("pay for play"), helped make it possible for new "multi-channel video programming" providers (MVPDs) to compete with cable operators, as did changes to the law to make it easier for competitors to get access to broadcast programming.[4] These policies of increasing competition were somewhat successful but their promise was not entirely fulfilled.[5] They enabled some new competitors to operate but these new competitors did not change the fundamental shape of the market. They did not slow the increasing power of cable generally and a few large cable companies in particular.[6] And they did little or nothing to keep the market from consolidating in ways detrimental to consumers and independent content producers alike. The Internet is changing the video marketplace just as it changed the market for music, news, books, and other forms of media. But it's not a foregone conclusion that the Internet will *disrupt* the video marketplace. Dominant players in the market today have control both over the content their nascent online competitors need for their service, and over the pipes they must use to reach consumers. As a result much high-value programming is not available online, and online video providers have to contend with artificially low bandwidth caps and other discriminatory practices that keep them from reaching their full potential.

Thus while it's inevitable that IP technologies and the Internet will play an ever-larger part of video delivery, it remains an open question whether consumers or incumbents with market power will enjoy the lion's share of the benefits. Consumers will still suffer from a lack of choice and independent content producers will still struggle to reach viewers if existing incumbents in the content and the MVPD industry continue to thwart disruptive change and manage the transition for their own benefit. Congress should once again take the necessary steps to ensure that incumbents cannot throttle (literally as well as figuratively) the legions of potential competitors trying to reach willing consumers.

At the same time, Congress should prune away the needless overgrowth of older rules, like syndicated exclusivity, the sports blackout rule and the network non-duplication rule, that exist only to protect the business model of local broadcasting. Senator DeMint and Representative Scalise are on the right track with their bill that would clear away much of the regulatory underbrush that holds back the evolution of the video marketplace,[7] although the bill goes too far by eliminating ownership restrictions still needed to maintain diversity in traditional media. Some other rules, like retransmission consent and the compulsory copyright license, are outdated, but part of an interwoven fabric of regulatory and business expectations. They should be reformed, but cautiously.

At the same time, measures that are designed to mitigate the market power of certain large video providers should not be repealed until true competition develops. In some respects they should be extended. For example, online video providers that wish to voluntarily operate as "multichannel video programming distributors" (MVPDs) under Title VI of the Communications Act8 should be able to do so, as this would enable them to access certain valuable content and protect them against anticompetitive actions by incumbents.[9] This would ensure that consumers had more choices for high-value content than they do today and would eliminate the incentives that keep certain content from being licensed widely.

Finally, the fact that the largest residential broadband ISPs, such as Comcast, are also MVPDs invested in the existing video distribution models raises concerns. These ISP/MVPDs can impose a variety of policies that prevent genuinely disruptive competition. For example, the ability to control how much data subscribers may access through data caps, the ability to privilege some content over others through prioritization or exemption from data caps, and the ability to control what devices can connect to the network, give cable operators (and other broadband providers like FIOS) the ability to pick winners and losers just as cable operators did from 1984 to 1992.

Detailed Analysis and Recommendations

The video marketplace is unique, not only because of its complicated business and regulatory structures, but because video incumbents are better placed to counter the threat the Internet poses to their business models. They do this in varied and creative ways.

Threats to Internet Openness

For a long time it looked as though ISPs would continue doing what Comcast did when it started degrading BitTorrent traffic—picking and choosing which Internet protocols and services got preferential or discriminatory treatment. But recently ISPs have found that it is more effective to discriminate via billing practices. Some ISPs have set their bandwidth caps so low as to make it financially unattractive to switch over entirely to online video, as this would put viewers over their caps and perhaps subject them to overage charges.[10] At the same time, at least one ISP exempts its own video services that are delivered over the same infrastructure from the caps.[11] These practices disadvantage services like Netflix and Amazon Instant Video and relegate most online video to the role of a supplement to, rather than replacement for, traditional MVPD services.

To counter this, Congress needs to stand behind the FCC's attempts to protect Internet openness.[12] At the same time these protections need to be strengthened, their loopholes need to be closed, and they need to take into account the fact that that discrimination can happen through billing, as well as through Internet "fast lanes" and other forms of technological discrimination.

Restrictions on the Availability of Content

The current regulatory system makes it so that incumbent MVPDs but not online providers can carry broadcast content,[13] and it makes it easy for incumbents to share content with each other while keeping it out of the hands of potential new competitors.[14] And while it's unlawful for incumbent providers to behave anti-competitively towards each other, they are free to keep their content away from online services, and to use exclusionary contracts and "most favored nation" clauses to limit the online distribution of independent programming.[15]

As a result, while a lot of very good video programming is available online, the most popular programming is not.[16] Popular broadcast and cable channels are not available online. Many popular shows are not available online at all or are only made available after a "windowing" period. Some programs are put online reasonably promptly, but are only viewable in inconvenient ways.

Some of the best online content is only available to viewers who also have cable subscriptions, through TV Everywhere and similar efforts. Live sports, and especially live local sports, are generally not available online at all. Thus, while online services make it easy to watch great documentaries, classic movies, and old sitcoms, the kinds of culturally-current programming that people talk about at the office and online are usually not available without a cable or satellite subscription.

This problem would be largely abated if online providers like Sky Angel were permitted to operate as MVPDs, like they want to.[17] The rules that ensure that all MVPDs can access certain content would then protect them as well as incumbents. At the same time, the FCC should find that the current rules that prohibit incumbents from behaving anti-competitively toward each other also prohibit them from taking anti-competitive acts against even those online video providers that choose not to operate as MVPDs.[18] But even short of that, if more content were available from online services that might choose to operate as MVPDs, the incentive to keep content offline would evaporate to the benefit of the entire video marketplace.

The current pay TV MVPD model is very lucrative for some because it forces viewers to pay for programming they don't want. Even some popular programmers like Time Warner, who have no direct stake in the cable business, find it more profitable to give exclusives to MVPDs than to make their programming available to willing buyers online.[19] This is because people pay for large bundles of cable channels, some of which are very expensive, even if they only want to watch a few.[20] Every cable subscriber has to pay for broadcast channels, even though they are available over the air for free. This leads to high prices that just keep getting higher. The result of all of this is a loss for consumers.

One quick way to fix this would be to scrap the rules that require that cable systems carry broadcast stations as part of their basic tier—customers should be able to choose what they pay for. And while video providers should be free to bundle content and should not be required to offer everything a la carte, it seems logical that increased competition from online providers would force today's providers to begin offering their customers more flexibility.

Marketplace Consolidation

The merger between Comcast and NBC Universal brought a large amount of programming under the control of a cable system that has an incentive to limit its distribution online.[21] While it is true that both the Department of Justice and the FCC conditioned their transaction on Comcast's commitment to make certain programming available to online distributors,[22] as Public Knowledge has argued, behavioral remedies are, in general, insufficient to overcome all the anti-competitive effects of mergers, joint ventures, and other structural changes that create incentives to limit distribution and innovation.[23] Unfortunately, yet another such change has been proposed, whereby Verizon and several large cable companies plan to create various joint entities to develop new video technologies and to market each other's products rather than compete.[24] In addition to limiting competition in existing markets, these arrangements could mean that much video in the future will be locked up in proprietary platforms, and could mean that anticompetitive "authentication" schemes like TV Everywhere become even more widespread. If policymakers truly wish to safeguard the future of video, they should prevent these sorts of anticompetitive agreements from taking place.

Outdated Rules That Protect Incumbent Business Models

Protectionist Measures

Finally, there are some rules on the books today that seem designed to prop up legacy business models and have long outlived any functions they may once have served.

Many of them can and should be repealed today. Examples of these include sports blackout rules, network nonduplication, and syndicated exclusivity provisions,[25] and the previously-mentioned rule that requires that all MVPD viewers pay for free over-the-air television.[26] Some of these rules were passed to protect aspects of the video distribution system from disruption before Internet video was a possibility, and when it seemed that if local broadcasters lost revenue nothing could replace them. Exclusivity rules not only keep cable systems from carrying signals from "distant" markets but they prevent networks from distributing content on a non-exclusive basis. The world these rules were written for is gone now and they have outlived their purpose. Some local broadcasters never provided unique local programming, and the various public goals that they provide can be achieved more effectively through other means.

Traditional models of video distribution are still valuable, and local broadcasters who serve their communities will continue to thrive after any regulatory reform. But the broadcasting industry no longer needs extraordinary protection against changes in technology, business models, and viewer behavior.

Outdated, but Complex Rules

Some other rules are outdated, but so interconnected with other rules and marketplace expectations that they need to be approached carefully. Among these are the compulsory copyright license,[27] retransmission consent,[28] and must-carry.[29] The compulsory license cannot be reformed unless video providers are given assurance that they never have to stop carrying programming just because they don't know whom to contact for a license, and to make sure that they can cope with any potential holdout problems. And it would make no sense to embark on a comprehensive reform of the laws governing video carriage in a way that replicated the problems that afflict the retransmission consent process today, while introducing new ones. Short of dealing with the compulsory license and retransmission consent together, several reforms could improve the current retransmission consent process. Many of the rules that have already been mentioned give an unfair advantage to broadcasters and drive up the rates they can charge.

And some broadcasters have engaged in brinksmanship tactics that harm viewers, where they pull their signals from MVPDs right before high-profile events.[30] These problems can at least be alleviated with meaningful "good faith" standards that discourage unfair negotiation tactics, and interim carriage requirements that minimize disruption to viewers.[31]

Finally, while the must-carry system is used by many low-value broadcasters in ways that Congress never intended, public and non-commercial stations continue to serve a valuable role and policymakers should find ways to protect the good that they do.

Policies That Are Still Needed

Still other rules serve a function and should be maintained, at least until competition develops. These include the program access, program carriage rules, as well as set-top box competition rules. The program access rules and related protections prevent any one MVPD from having exclusive rights to content. Although the video market is not as competitive as it can be in the Internet age, the fact remains that the American video distribution market is more competitive than that of many other countries.

The program access rules are to thank for that, and they should be extended to all services that wish to operate as MVPDs, even ones that are exclusively online.

Similarly, the program carriage system, which protects independent programmers from the negative effects of bottleneck control by some MVPDs, still serves a role in ensuring that viewers can enjoy content from diverse sources. Lastly, the directive expressed by Congress under Section 629 to have true set-top box competition has remained largely unfulfilled. Until Internet-delivered video becomes a true substitution, preserving the FCC's authority to promote set-top box competition will remain necessary.

Copyright and Spectrum Policy

There are two other kinds of regulations that can hold back the development of online video. Policymakers don't always see them as "regulations" in the same sense as things like syndicated exclusivity. But copyright and spectrum laws are regulations nonetheless, and they have profound effects on the shape of the market.

Copyright law shouldn't be misused to hold back the evolution of the video marketplace. Dish is being sued for making a DVR that's too sophisticated for the taste of some networks. But it's not illegal to skip commercials or for users to take full advantage of their home recording rights.[32] And Aereo's remote antenna is legal just as Cablevision's remote DVR is.[33] Copyrights are limited monopolies granted by the government, and they come with a series of limitations and exceptions designed to protect users as well as creators. They should not be a weapon used to limit experimentation with business models and services.

Nor should misplaced fears of piracy keep content offline. Some content industry executives have a view of technology and the Internet that can only be described as superstitious, and they think that if they give people access to content they'll lose control of it. But recent history shows that most people only turn to piracy when content is not available online though other means. From the perspective of limiting copyright infringement, limiting online distribution is simply counterproductive.

This is why it is particularly distressing that recent trade agreements contain language that could be interpreted as limiting the possibilities of online video distribution.[34]

To whom the government assigns spectrum and for what purpose it allocates it also has an impact on the video marketplace. As long as broadcasters use the public airwaves they will have public responsibilities. For example, they must operate transparently,[35] they must serve the needs of their communities,[36] and they cannot behave unreasonably in retransmission negotiations.[37] While it is true that fewer people rely on over-the-air television today than they did in its peak, due to the increasing costs of cable, a new generation of viewers is getting familiar with rabbit ears.38 Thus, to say that broadcasting is no longer relevant is just as wrong as to say that it should remain at the center of the video marketplace. In a more competitive video marketplace there will no doubt be room for many different kinds of services. The solution is not to enshrine or attack broadcasting but to incentivize them to create great content, and to adopt policies that allow spectrum to be put to other uses. Not only would this be beneficial to communications policy generally but the impact on the video marketplace would be profound, as distribution channels adapt to fit a more decentralized and dynamic marketplace.

CONCLUSION

As they have in the past policymakers are starting to consider the implications of increasing change in the market for video distribution. History provides examples both of protectionist regulations that should be avoided today, and of pro-competitive measures that serve as more positive precedents. But today is different in one way: Finally, the technology exists that could eliminate the physical, bottleneck control of video distribution that has existed in various forms for decades. If policymakers take some simple steps to facilitate the development of competitive online video they can begin to disengage from regulations that were designed to counter the effects of this bottleneck control. However, if they fail to do this, it is likely that incumbents will be able to continue to shape the development of the video market and extend their current dominance indefinitely. While the Internet provides grounds for hoping that the future of video will be better for consumers, policymakers have a lot of work to do to help make that happen.

End Notes

[1] *See United States v. Southwestern Cable Co.*, 392 US 157 (1968). This case, in addition to being an important case setting out the bounds of FCC authority, contains a summary of the FCC's early efforts at cable regulation. *See also* OFFICE OF TELECOMMUNICATIONS POLICY, CABLE: REPORT TO THE PRESIDENT (1974), which contains an early history of the cable industry and attempts at cable regulation, as well as policy recommendations.

[2] The 1974 OSTP Report said that "cable is not merely an extension or improvement of broadcast television. It has the potential to become an important and entirely new communications medium, open while and available to all." OSTP Report at 13. But cable did succeed in providing viewers with more content it fell short of this early promise, and the regulatory system that developed ensured that cable extending the reach of broadcasting instead of developing into a competitor to it.

[3] Cable Television Consumer Protection and Competition Act of 1992, PL 102-385, 106 Stat. 1460 (1992).

[4] E.g., The Satellite Home Viewer Improvement Act of 1999, Pub.L. No. 106-113, 113 Stat. 1501, 1501A-526 to 1501A-545 (Nov. 29, 1999).

[5] *See* Annual Assessment of the Status of Competition in the Market for Video Programming, Thirteenth Annual Report, MB Docket No. 06-189 (rel. Jan 16, 2009), available at http://

hraunfoss.fcc.gov/edocs_public/attachmatch/FCC-07-206A1.pdf. *See also* Comments of Public Knowledge in Annual Assessment of the Status of Competition in the Market for the Delivery of Video Programming, MB Docket No. 07-269 (June 8, 2011), *available at* http://www.publicknowledge.org/files/docs/PK_Comments_MVPD-Competition-Report.pdf.

[6] For example, Adelphia's cable assets were sold to Time Warner Cable and Comcast. *See* Adelphia Sold to Time Warner, Comcast, BUFFALO BUSINESS FIRST, April 21, 2005, http://www.bizjournals.com/buffalo/stories/2005/04/18/daily37.html?page=all. Comcast's cable assets and NBC Universal have been combined in a joint venture that is controlled by, and 51% owned by Comcast. *See* General Electric, New NBCU, http://www.ge.com/newnbcu.

[7] Next Generation Television Marketplace Act, H.R. 3675 and S. 2008.

[8] 47 USC §602 (13)

[9] *See* Comments of Public Knowledge in Interpretation of the Terms "Multichannel Video Programming Distributor" and "Channel" as Raised in Pending Program Access Proceeding, MB Docket No. 12-83 (filed May 14, 2012) (Sky Angel Comments), *available at* http://www.publicknowledge.org/interpretation-mvpd.

[10] ANDREW ODLYZKO, BILL ST. ARNAUD, ERIK STALLMAN, & MICHAEL WEINBERG, KNOW YOUR LIMITS: CONSIDERING THE ROLE OF DATA CAPS AND USAGE BASED BILLING IN INTERNET ACCESS SERVICE 48 (Public Knowledge 2012) ("Comcast's own estimate for the amount of data required to replace its pay-television offering with an over the top competitor is 288 GB per month. In light of this, it may come as no surprise that Comcast's data cap is set at 250 GB per month."). Comcast has since raised its cap, but it is worth observing that the 288 GB per month figure is based on an unknown mix of standard and high-definition content; presumably, a higher percentage of high-definition video would lead to a higher figure. *See* Mark Israel and Michael L. Katz, *The Comcast/NBCU Transaction and Online Video Distribution*, Submitted by Comcast Corporation, MB Docket No. 10-56 (May 4 2010) at 33, *available at* http://apps.fcc.gov/ecfs/document/view?id=7020448237.

[11] Michael Weinberg, *Comcast Exempts Itself From Its Data Cap, Violates (at least the) Spirit of Net Neutrality*, PUBLIC KNOWLEDGE (March 26, 2012), http://www.publicknowledge.org/blog/comcast-exempts-itself-its-data-cap-violates-.

[12] Preserving the Open Internet, *Report & Order*, GN Docket No. 09-191, FCC 10-201, *available at* http://fjallfoss.fcc.gov/edocs_public/attachmatch/FCC-10-201A1.pdf.

[13] 47 U.S.C. § 325; 47 C.F.R. § 76.64.

[14] 47 U.S.C. § 548. *See* Revision of the Commission's Program Access Rules, *Notice of Proposed Rulemaking*, MB Docket No. 12-68, FCC 12-30 (rel. Mar. 20, 2012) for an overview of the program access rules.

[15] Jon Brodkin, *DOJ probing Big Cable Over Online Video Competition*, ARS TECHNICA, June 13, 2012 (noting that "[t]he DOJ is also investigating contracts programmers sign to be distributed on cable systems, which include 'most-favored nation clauses' that may favor cable companies over online video distributors.")

[16] *See* Carlos Kirjner, Internet TV (or Why It Is So Hard to Go Over the Top), Bernstein Research (June 15, 2012).

[17] *See* Sky Angel Comments.

[18] As Public Knowledge has argued,

> The [FCC] should use its authority over the video programming distribution market to protect online video distribution generally, by prohibiting MVPDs from behaving anticompetitively in ways that harm any video distributor, whether or not it is an MVPD. Section 628 of the Communications Act provides authority for this. This Section bans any actions "the purpose or effect of which is to hinder significantly or to prevent any multichannel video programming distributor from providing ... programming to subscribers or consumers." The close connection between the markets for MVPD and non-MVPD video distribution mean that anticompetitive actions taken against an non-MVPD would likely have a deleterious effect on the ability of a competitive MVPD to offer programming—for

example, by increasing its costs, or inhibiting the ability of an MVPD to offer programming on demand or online.

Sky Angel Comments at 24-25.

[19] Brian Stelter, *HBO Says No, for Now, to Fans Who Want a Web-Only Option*, NY TIMES MEDIA DECODER, June 6, 2012, http://mediadecoder.blogs.nytimes.com/2012/06/06/hbo-says-no-for-now-to-fans-who-want-a-web-only-option.

[20] Peter Kafka, *Hate Paying for Cable? Here's Why*, ALLTHINGSD, March 10, 2010, http://allthingsd.com/20100308/hate-paying-for-cable-heres-the-reason-why.

[21] Competitive Impact Statement of the Department of Justice at 4, United States v. Comcast Corp., 1:11-cv-00106, (D.D.C. Jan. 18, 2011), available at http://www.justice.gov/atr/cases/f266100/266158.pdf

[22] Applications of Comcast Corporation, General Electric Company and NBC Universal, Inc., for Consent to Assign Licenses and Transfer Control of Licenses, Memorandum Opinion & Order, 26 FCC Rcd. 4238 (2011); Final Judgment in United States v. Comcast, United States District Court for the District of Columbia, Case No. 1:II-cv-00I06 (Sept. 1, 2011).

[23] Petition to Deny of Public Knowledge and Future of Music Coalition in WT Docket No. 11-65 (filed May 31, 2011), at 62-70, available at http://www.publicknowledge.org/files/docs/pk_fmc-att_tmo-petition_to_deny.pdf.

[24] *See* Petition to Deny of Public Knowledge et al. in WT Docket No. 12-4 (filed Feb. 21, 2012), *available at* http://www.publicknowledge.org/files/pk_verizon_spectrumco_petition.pdf.

[25] 47 C.F.R. §§ 76.92(f), 76.106(a), 76.111, 76.120, and 76.127-130.

[26] 47 C.F.R. § 76.901(a) ("The basic service tier shall, at a minimum, include all signals of domestic television broadcast stations provided to any subscriber"); 47 C.F.R. § 76.920 ("Every subscriber of a cable system must subscribe to the basic tier in order to subscribe to any other tier of video programming or to purchase any other video programming.").

[27] 17 U.S.C. § 111.

[28] 47 U.S.C. § 325; 47 C.F.R. § 76.64.

[29] 47 U.S.C. § 534; 47 C.F.R. § 76.55.

[30] Amendment of the Commission's Rules Related to Retransmission Consent, *Notice of Proposed Rulemaking*, 26 FCC Rcd. 2718, ¶15 (2011).

[31] *See* Comments of Public Knowledge and New America Foundation in MB Docket No. 10-71 (filed May 27, 2011), *available at* http://www.publicknowledge.org/files/docs/11-05-27PK-NAF_retrans_comments.pdf.

[32] *See* John Bergmayer, *Networks Pull the Trigger on Dish, but They're Only Hurting Themselves*, PUBLIC KNOWLEDGE (May 25, 2012), http://www.publicknowledge.org/blog/networks-pull-trigger-against-dish-theyre-onl.

[33] *See* Brief of Amici Curiae Electronic Frontier Foundation and Public Knowledge in *WNET v. Aereo*, United States District Court for the Southern District of New York, Case No. 1:12-cv-01543-AJN, *available at* http://www.publicknowledge.org/files/aereo_amici_brief.pdf.

[34] Many free trade agreements appear to state that online retransmission may not occur without the permission both of the owner of the copyright in the programming, and of the broadcaster. This is at odds with the current system of a compulsory license plus retransmission consent, which requires MVPDs to obtain the permission only of the signal owner, not of the content owners. Some current reform proposals involve requiring an MVPD to obtain the permission of the copyright holders instead of the permission of the broadcaster, but not of both. *See* John Bergmayer, *The US-Colombia Free Trade Agreement: Policy Laundering in Action*, PUBLIC KNOWLEDGE (April 20, 2012), http://www.publicknowledge.org/blog/us-colombia-laundering.

[35] *See* 47 C.F.R. § 73.3526 (public file requirement).

[36] 47 C.F.R. § 73.3526(e)(11)(i).

[37] 47 U.S.C. §§ 325(b)(3)(C)(ii), (iii).

[38] Christopher S. Stewart, *Over-the-Air TV Catches Second Wind, Aided by the Web*, WALL STREET JOURNAL (Feb. 21, 2012), http://online.wsj.com/article/ SB10001424052970204059804577229451364593094.html ("It's cool to have rabbit ears again.").

In: Future of Audio and Video Markets ... ISBN 978-1-62257-764-4
Editors: C. Powell Long, F. McDonald

Chapter 7

STATEMENT OF DAVID BARRETT, PRESIDENT AND CEO, HEARST TELEVISION. HEARING ON "THE FUTURE OF VIDEO"*

Good morning, Chairman Walden, Ranking Member Eshoo, and members of the Subcommittee, and thank you for inviting me to testify today. My name is David Barrett, and I am President and CEO of Hearst Television, Incorporated, which operates 29 television stations across the U.S. I am testifying today on behalf of the free, local, over-the-air television members of the National Association of Broadcasters.

I. INTRODUCTION

In my view, the future of broadcast video services is bright. Recent data show that the number of viewers accessing television over the air (OTA) has increased dramatically in recent years. Results of a survey released last week by GfK Media/Knowledge Networks show that about 21 million households, representing 54 million viewers, now access digital broadcast television exclusively through an antenna.[1] That is a sharp increase – nearly 20 percent – over just a few years ago. And nearly a quarter (24 percent) of homes headed by younger adults, those with a head of household aged 18-34, rely on over-the-air reception for their broadcast television viewing. *Id.*

---

* This is an edited, reformatted and augmented version of the testimony given on June 27, 2012 before the House Committee on Energy and Commerce.

Who are these viewers? In addition to young people, many are low-income families or minorities. *Id.* The GfK Media report shows that the effects of the economic downturn, increasing subscriber fees for cable and satellite TV, and the plethora of new broadcast options in the digital age have led many consumers to embrace broadcast TV again. This is exciting news for broadcasters and should inform Congress as it oversees the Federal Communication Commission's (FCC) management of spectrum allocated to free, over-the-air television.

Some of this resurgence can also be attributed to technological advances in broadcast TV. The television industry recently passed the three-year anniversary of the transition to all-digital distribution. By almost any measure, the transition and broadcasters' embrace of digital technology have been a tremendous success and a boon for viewers. Nearly every major television broadcaster now provides its content to viewers in crystal-clear high definition over the air for free. Most stations also offer anywhere from one to up to three additional "multicast channels" – extra channels containing new and diverse program content, and all of which operate in the same 6 MHz of spectrum that previously held just one analog channel. This new, free, digital over-the-air service doubles, and, in some cases, more than triples, the number of channels available. Indeed, broadcasters' ability to multicast has led to the rise of multiple new national networks, including many networks, such as Bounce TV, Estrella, Live Well, and MeTV, that serve more specialized, diverse, and ethnic audiences. This trend will continue as new networks grow their audiences with increasingly diverse and compelling programming.

With these developments in mind, this testimony first addresses spectrum and broadcasting's role in the communications ecosystem. It then focuses on important issues regarding the "rules of the road" for video services.

With regard to spectrum, NAB urges Congress to remain vigilant in its oversight of the process of broadcast incentive auctions. Incentive auctions themselves are unprecedented, and the television spectrum auction specifically will have a direct impact on millions of viewers, potentially exceeding that of the digital TV transition. It is critical that this Committee ensure the FCC implements incentive auctions consistent with statutory requirements and Congressional intent.

Beyond these auctions, we should also be focused on the future of broadcasting and how it can, and should, play a vital role in our nation's communications system moving forward. Beyond continuing to serve viewing audiences and local communities as we always have, the broadcast industry's devolving technology will be a critical complement to wireless broadband.

Just as wireless companies are upgrading their technology, from 3G to LTE and beyond, broadcasters will also be upgrading, and the results could have an extraordinary impact on spectral efficiency.

My testimony today also responds to continuing calls by pay television services to revise the legal framework of video programming distribution. These companies would have Congress change the laws and regulations that have successfully governed the video marketplace for decades. They would turn back the clock to days when broadcasters were essentially forced to subsidize their pay TV competitors. As explained below, such efforts are contrary to the public interest and should be rejected.

II. Congress Should Ensure That Incentive Auctions Are Implemented as Intended

A. Transparency Is Critical to Incentive Auction Success

Earlier this year, as part of the Middle Class Tax Relief and Job Creation Act of 2012 (Public Law 112-96) ("Incentive Auction Act"), Congress gave the FCC the authority, for the first time, to conduct incentive auctions, including auctions of broadcast spectrum. While we dispute some of the underlying orthodoxy behind the push for incentive auctions – namely, that repurposing large amounts of broadcast spectrum is necessary to solve a looming spectrum "crisis" – we nonetheless supported the legislation after this Committee and Congress included several appropriate viewer safeguards in the legislation. These provisions ensure that the auctions are completely voluntary and that the Commission must make every effort, after the television auction and the repacking of stations into a smaller band, to replicate the service areas of the many stations that will remain on the air serving their local communities. With digital broadcasting, even a seemingly slight reduction in television stations' service areas could result in a loss of service for millions of viewers. And this could have a particularly big impact on rural viewers. I encourage this Committee to ensure that the FCC follows the intent of Congress in this important respect.

The success of incentive auctions ultimately will be defined by their results for the American people. The Commission must maintain a robust broadcasting system that continues to provide free and local television service to millions of viewers, while moving to provide a strong and fast wireless broadband system. To achieve this result, the Commission must fully engage all the affected industries. For broadcasters, that includes not just those stations that may choose to participate in the auction, but also those stations that do not – yet will nonetheless be moved, or "repacked," to a new channel.

Broadcasters are understandably apprehensive about this process. Their concerns include how many stations will be moved and whether there will be enough remaining channels to accommodate those stations that wish to continue to serve the public. To ease that apprehension, the Commission should be as transparent as possible about how it plans to conduct incentive auctions and how it plans to repack stations. NAB is concerned that some well-intended language in the Incentive Auction Act ensuring confidentiality for broadcasters that submit offers to sell in the reverse auction could be interpreted in a way that would undermine this goal.[2] Careful implementation of this provision is important.

While in some respects confidentiality is a good idea, that confidentiality should not extend to the operational mechanisms of the incentive auction and the repacking process. Broadcasters should know, for example, what modeling and service area assumptions the Commission makes as it lays out a repacking plan, how the Commission will coordinate with Canada and Mexico (which potentially affects hundreds of stations near the borders), and what impact the repacking process will have on existing station coverage areas. Transparency benefits broadcasters, potential bidders in the auction, and the FCC. Broadcasters are more likely to submit an offer to sell in the reverse auction or to consider the channel-sharing option if they understand fully how the process will work *before the incentive auction begins.* Limiting the release of information about the mechanics of the incentive auction process will increase the likelihood that the incentive auction will not be successful.

B. The One-to-Many Broadcasting Model Is a Necessary Complement to the One-to-One Broadband Model, Now and in the Future

One of the most important, but least understood, reasons for maintaining a robust broadcasting system is the critical role that broadcasters play in the wider communications ecosystem. Broadcasting's one-to-many video and data service, plainly, is the most spectrally efficient wireless delivery system for high demand content. As consumers rely on handheld devices – including mobile phones and tablets – to access content like video, the one-to-many broadcasting model will become increasingly *more important*, not less important. For example, mobile digital TV, which is currently being rolled out by broadcast stations around the country, can alleviate pressure on wireless networks when viewers using mobile devices access popular television programming, such as sporting events, because consumers will not have to access that content through the cellular network. More importantly, as evidenced by the lifesaving role the technology played in the Japanese earthquake last year, mobile digital TV is the best way to reach on-the-go viewers with critical information. Future broadcast distribution standards being developed now will be able to deliver not just high-demand video, but also high-demand data of all types, in the most spectrally efficient manner, greatly reducing the burden on over-taxed cellular networks.

In considering what our nation's communications system should look like in 10 or 20 years, it is critical that we avoid any policy relying too heavily on a one-to-one architecture to the detriment of the diverse broadcast model. The point-to-point architecture of wireless broadband networks essentially means that each user has his or her own path in the cellular network. This type of design allows two people standing next to each other using the same type of device and operating on the same wireless network to access totally different types of information.

The first person can be watching a video and the second person can be looking up directions to the closest restaurant. But, if those two people and hundreds or thousands of other people near them are trying to access the same information at the same time – which occurs during emergencies – the wireless network will quickly be overwhelmed.

In contrast, television and radio broadcasting creates one or just a few data streams and transmits that data over a specific geographic area using a high-powered transmitter. This data can be received by anyone who has a receiver located within the transmission range of that broadcaster. Adding more users to the broadcast distribution network has no impact on its ability to deliver information. For high demand information, this is the ideal distribution method.

As broadcast technology continues to evolve, we expect that the broadcasting model will work cooperatively and seamlessly with cellular networks to deliver information to wireless devices, delivering, as broadcasters do today, the content that consumers seek the most. In addition, broadcast licenses are held by many diverse licensees. The broadcast model permits a diversity of ownership and control that does not exist in wireless services.

This basic broadcast-broadband wireless model will clearly benefit consumers, who will face higher wireless bills and caps on data usage in the next few years. Already, most major wireless companies are eliminating their unlimited data plans. And as the transition to 4G wireless technology has shown, consumers quickly exceed their data limits, and pay exorbitant overage fees, when accessing video through cellular networks.[3]

Coupling a high-powered one-to-many broadcast transmission with those cellular networks would eliminate this concern for consumers accessing popular video content.

C. There Is Still a Need for a Comprehensive Spectrum Inventory

I would also like to thank Chairman Walden and this Committee for initiating a new spectrum task force, focused on finding ways to analyze and optimize use of the vast quantities of spectrum under federal government control. To that end, it is critical for the task force to have a complete picture of how all spectrum, both federally controlled and commercial, is being used, including a clear understanding of who currently holds spectrum, how they are using it, and the intensity and effectiveness of their use. Such an inventory of spectrum deployment should not delay the incentive auction process, but an inventory is prudent and necessary for Congress, the National Telecommunications and Information Administration (NTIA), and the FCC to make informed decisions on U.S. spectrum policy.

The wireless industry claims there is a looming spectrum crisis, yet the industry appears to be warehousing vast quantities of unused spectrum.[4] If the wireless industry were truly running out of bandwidth, it already would have developed more of the spectrum that it currently controls. There have been a number of industry analysts who have cast doubt on the claims of a spectrum crisis. Martin Cooper, the father of the cell phone, was recently quoted in a *New York Times* article claiming that available technologies, which could greatly increase spectrum efficiency, are not being implemented by the wireless industry.[5]

In short, NAB supports Congress' decision to grant the FCC authority to conduct voluntary incentive auctions of broadcast spectrum as long as the Commission fully implements the Incentive Auction Act's viewer safeguards and protects the millions of Americans who rely on free, over-the-air local broadcast stations.

III. The Current Legal Framework Governing Broadcaster-Pay TV Relationships Serves the Public Interest

Turning to the challenges facing broadcast video services, television broadcasters offer a high quality, free, over-the-air, locally-oriented service that competes head-to-head with nationally-oriented pay TV platforms, hundreds of non-broadcast subscription networks, and other numerous programming sources.

Congress already has in place laws that successfully govern the relationship between pay TV providers and broadcasters. These laws have a single purpose. They are designed to assure fair competition in a highly competitive media market and maximize the diversity, quality, and affordability of television service to the American people.

This legal framework works because it serves the needs of television viewers and reflects the actual business relationships between broadcasters and pay TV providers.

Two bills currently before Congress, H.R.3675 and S.2008, both known as "The Next Generation Television Marketplace Act of 2011," will harm local stations and television viewers in at least three ways.

First, they would turn back the clock to a time when cable and satellite providers confiscated and resold broadcast signals to their subscribers without obtaining broadcasters' consent — a time when broadcasters were forced to subsidize their pay TV competitors.

Aside from being fundamentally unfair, a return to this system would seriously threaten broadcasters' ability to invest in high quality informational and entertainment programming to serve viewers and to compete effectively for audiences and advertisers.

Second, the bills would eliminate the ability of the FCC to enforce privately negotiated contracts between program distributors and stations for the distribution of network and syndicated programming on an exclusive basis.

Finally, these bills would further compound the harm to consumers by eliminating statutory provisions that promote fair competition between pay TV providers and the free, over-the-air broadcast service. And they would impair the ability of *any* regulatory body to protect consumers from escalating pay TV bills.

A. Congress Should Not Change Its Well-Functioning System of Retransmission Consent by Tilting the Marketplace in Pay TV Providers' Favor

One point that is sometimes lost in discussions of retransmission consent is why Congress granted broadcasters retransmission rights in the first instance. In short, Congress adopted retransmission consent in 1992 to ensure that broadcasters had the opportunity to negotiate at arm's length in the marketplace for compensation in exchange for the right of cable and other multichannel video programming distributors to resell their broadcast signals. This law promotes fair competition in the video marketplace, is pro-consumer, and enhances the vibrancy of the nation's free, over-the-air broadcast service, as Congress intended. It also benefits television viewers – your constituents – in markets across the country by assuring free access to vital news, emergency and weather information, public service programming, and a variety of entertainment and sports programming. This service is of special significance and importance to your constituents who cannot afford an expensive pay TV subscription service. These policy goals remain just as important today as when they were enacted.

Prior to the Cable Television Consumer Protection and Competition Act of 1992 (the "1992 Act"), cable operators were not required to seek the permission of a station before retransmitting and reselling its signal, nor were they required to negotiate with the station for that privilege. At a time when cable systems had few channels and were limited to an antenna function of improving the reception of certain local broadcast signals, this lack of recognition for the rights broadcasters possess in their signals had limited practical significance. However, the video marketplace changed dramatically in the 1970s and 1980s. Cable systems began to include not only local signals, but also distant, duplicating broadcast signals and the programming of vertically-integrated cable networks and premium services. Cable systems began to compete head-to-head with broadcasters for viewers and for national and local advertising revenues, but they were still allowed to resell local broadcast signals to their paying subscribers without the permission or consent of the station.

By the early 1990s, Congress concluded that this failure to recognize broadcasters' rights in their signals had "created a distortion in the video marketplace" that "threaten[ed] the future of over-the-air broadcasting." S. Rep. No. 92, 102d Cong., 1st Sess. at 35 (1991) (*Senate Report*). Using the revenues they obtained from reselling broadcast signals, cable systems had supported the creation of cable programming (including program networks vertically integrated with cable system operators) and were able to sell advertising on these cable channels in direct competition with broadcasters.

Given this dramatic change in the nature of cable systems, program services and advertising practices, Congress determined that the then-existing law was not only unfair to local broadcast stations, it was anticompetitive.

Specifically, Congress concluded that public policy should not support "a system under which broadcasters in effect subsidize the establishment of their chief competitors." *Id.* Noting the continued popularity of broadcast programming, Congress also found that a very substantial portion of the fees that consumers pay to cable systems is attributable to the value they receive from watching broadcast signals. *Id.* To remedy this "distortion," Congress in the 1992 Act gave broadcasters control over the use of their signals and permitted broadcasters to seek compensation from cable operators and other multichannel video programming distributors (MVPDs) for carriage of their signals. *See* 47 U.S.C. § 325(b). Congress specifically noted that cable operators pay for the cable programming they offer to customers and that programming services originating on broadcast channels and resold by MVPDs should be treated no differently. *Senate Report* at 35.

In establishing retransmission consent, Congress intended to create a "marketplace for the disposition of the rights to retransmit broadcast signals." *Id.* at 36. Congress emphasized that it did not intend "to dictate the outcome of the ensuing marketplace negotiations" between broadcasters and MVPDs. *Id.* Retransmission consent does not guarantee that a broadcaster will receive fair compensation from an MVPD for retransmission of its signal; it only provides a broadcaster with an opportunity to negotiate for compensation of various types.

The reasons for establishing this retransmission marketplace remain as valid and important today as they were in 1992. Congress enacted retransmission consent because it recognized the value of broadcasters' signals, which continue to be highly valued by viewers and advertisers today.[6] It is still the case today that pay TV providers would like to confiscate the signals of local broadcasters – their competitors for viewers and advertisers – and resell those signals to paying subscribers. But it would be as unfair and anticompetitive today to allow pay TV operators to do this without the consent of local stations as when the retransmission consent statute was enacted. Moreover, from the inception of the Radio Act of 1927, one broadcast station has been unable to take the signal of another broadcast station without the originating station's consent. Similarly, a broadcast station cannot intercept a cable system's transmission or a satellite carrier's signal and then rebroadcast it without consent. It would be the height of unfairness to single out broadcast stations for such disparate and anticompetitive treatment by allowing MVPDs – their competitors – to confiscate their signals without authorization.

It is still true today that cable and satellite operators pay for all of the other, non-broadcast programming they offer to attract subscribers (and, in fact, pay *more* for that programming on a per viewer basis). And there is still no reason that broadcasters should be uniquely disfavored and not be allowed to negotiate for others' use of their signals. In sum, Congress's original goals of correcting distortions in the video marketplace, promoting competition, and "ensur[ing] that our system of free broadcasting remains vibrant," continue to be served today by the retransmission consent system. S*enate Report* at 36.

While pay TV providers contend that changes in the marketplace since 1992 have somehow undermined the rationale for retransmission consent, they focus exclusively on the advent of limited competition in the MVPD market and disregard other changes that benefit MVPDs in retransmission negotiations. For example, a much larger percentage of television viewers subscribe to pay TV services today than in 1992. Because broadcasters rely very heavily on advertising revenue, and, therefore, seek to reach the largest viewing audience possible, broadcasters today have a stronger incentive than ever to conclude retransmission negotiations successfully and avoid carriage disputes, which result in the immediate loss of both retransmission consent compensation and advertising revenues for local stations. Moreover, the MVPD market also has grown increasingly consolidated over time, with just ten MVPDs serving 90% of pay TV subscribers nationally, and with a majority of cable subscribers served by systems that are part of regional cable system "clusters." As a result, local broadcasters (including small to medium-sized stations and groups) often must deal with powerful, consolidated, and highly-concentrated and vertically-integrated MVPDs in retransmission consent negotiations. These consolidated MVPDs increasingly compete with broadcasters for viewers and for national and local advertising revenues, thereby additionally fragmenting local stations' audiences and advertising revenues. Congress should refrain from intervening in the retransmission consent process at the behest of these pay TV providers.

Beyond tilting the retransmission consent marketplace more in the favor of pay TV providers, changes to the retransmission consent system are entirely unnecessary and would be contrary to the interests of consumers.

Pay TV providers sometimes assert that they are paying retransmission consent fees that represent dramatic increases from one year to the next. Such percentage descriptions of increases may seem significant until you consider that the increase may be from little or no compensation to a small amount of compensation (e.g., an increase from one cent per subscriber per month to two cents per subscriber per month is a 100% increase, but is a very small amount). Retransmission consent compensation to broadcasters represents but a tiny fraction of what MVPDs spend on other programming and what they earn in revenues. For example, in 2010, retransmission consent fees were only about six-tenths of one percent of cable industry revenues.[7] Data recently published by SNL Kagan show that, in 2011, retransmission consent fees represented a total of 1.46 billion dollars, compared to 26.66 billion dollars paid for basic cable networks.[8] Most of these networks have considerably lower ratings than broadcast television stations.[9]

In addition, the vast majority of retransmission consent agreements are successfully negotiated without disruption of any kind to customers of pay TV providers.

NAB has repeatedly commissioned studies of negotiating impasses and their impact on television viewers. These studies have shown that consumers are over 20 times more likely to be deprived of television viewing by an electricity outage than by a bargaining impasse between broadcasters and MVPDs. From 2006 - 2011, aggregate service interruptions from retransmission consent negotiating impasses represented approximately *one one-hundredth of one percent* of annual U.S. television viewing hours.[10] Moreover, no broadcaster – not a single one – has ever been found by the FCC to have breached its obligation to negotiate retransmission consent in good faith. The same, however, cannot be said of pay TV companies.[11] Especially in light of pay TV providers' continual increases in the rates they charge consumers – increases consistently above the rate of inflation – policymakers should disregard these providers' factually unsupportable attempts to characterize their attacks on retransmission consent as protecting the interests of consumers.

Economic studies have shown that curtailing local stations' ability to obtain retransmission consent revenues would significantly reduce investment returns in the broadcast industry and reduce the amount of local news, public service, and public safety programming produced by stations.[12] In light of the economic challenges facing all providers of local journalism, Congress should refrain from undermining the retransmission consent system that increasingly supports local broadcast journalism important to viewers who subscribe to pay TV services, as well as to the growing numbers who receive all television over the air.

B. The Program Exclusivity Rules Work in Tandem with Retransmission Consent to Protect Localism, Diversity, and Private Contract Rights

Through arms'-length free market negotiations with program providers (including networks and syndicators), local broadcast stations purchase and pay for the exclusive rights to carry certain programming within a limited geographic area. The FCC's network non-duplication and syndicated exclusivity rules *do not* impose program exclusivity – they simply allow the FCC to enforce privately negotiated program exclusivity agreements between program suppliers and broadcast stations. In fact, the FCC's rules actually limit and restrict the geographic area of exclusivity to assure fair competition between stations and between cable systems. Both H.R.3675 and S.2008 would require the FCC to eliminate its program exclusivity rules.[13] While cable and satellite interests have sought to paint these rules as regulatory "protections," it is useful to take a closer look at what these rules really entail, why they exist, why they are important, and how they, like retransmission consent, promote competition in the creation and distribution of television programming.

A fact often missed in debate over the network non-duplication and syndicated exclusivity rules is that the rules themselves *do not* provide program exclusivity. In fact, the rules actually *limit and restrict* program exclusivity by limiting the geographic area in which television stations may enter into program exclusivity agreements with network and syndicated program suppliers.[14] The actual program exclusivity terms for network non-duplication and syndicated program exclusivity are a matter of *private contractual agreement* between the program supplier and the local television station. Neither the FCC nor its rules provide or enforce program exclusivity provisions or arrangements not agreed to by the program supplier and the local station. The reality is that, subject only to antitrust law, in the *absence* of the FCC's network non-duplication and syndicated exclusivity rules, program suppliers and local television stations could enter into exclusivity arrangements covering geographic areas of hundreds of miles.

In attacking the FCC's rules, what cable and satellite operators actually want is the adoption of mandatory "broadcast signal access" rules in abrogation of market-based, freely-negotiated program contracts – all for the single purpose of securing an unfair, government-granted competitive advantage over local television stations. For example, MVPDs complain that they are limited by program exclusivity contracts in obtaining duplicative broadcast programming from other sources outside the local market. In fact, that is the whole point of exclusivity contracts – exclusivity is valued in the marketplace and ultimately induces the provision of greater programming choice and quality for consumers, as further explained below. MVPDs also complain that the program exclusivity rules somehow confer an unfair advantage on broadcast stations, conveniently ignoring the advantage that MVPDs would otherwise have in exercising their *own* freedom to enter into exclusive programming contracts (a notable example being DIRECTV's NFL Sunday Ticket).

Exclusivity – as Congress and the FCC have consistently recognized – constitutes an essential component of America's unique system of free, over-the-air television stations licensed to serve local communities.[15] Local affiliates always have negotiated with networks and syndicated programming sources for exclusive programming within their markets. Advertisers on local broadcast stations expect and, indeed, pay for that exclusivity; these advertising revenues support stations' local programming, including news, and their ability to serve their communities. Exclusivity, which is limited by FCC rules to narrowly defined geographic zones near stations' home communities, enhances competition by strengthening local stations' ability to compete against the hundreds of non-broadcast and non-local programming networks offered by cable and satellite. As noted above, the FCC's rules do not mandate exclusivity, but merely enable broadcasters to protect the contractual arrangements they have entered into for the very purpose of securing programming content that meets the needs and interests of their communities.

Program exclusivity, and the system of local service it permits, thus is not a weakness of our broadcast system, as MVPDs often claim. It is a unique and highly valued strength. As with retransmission consent, there is no warrant for additional government intrusion into this realm of purely private contractual negotiations. As the FCC concluded when it examined its exclusivity rules in detail, interference into the contractual relations between broadcasters, networks and syndicated programming suppliers would "contradict our own requirements of broadcast licensees and would hinder our policy goals."[16]

C. Eliminating Mandatory Carriage Would Further Harm Viewers

As part of the 1992 Act, Congress found that cable operators had the incentive and ability to favor their own programming over the programming of competitors, including local television broadcast stations.[17] The must carry provisions of the 1992 Act were based on a finding that action was necessary to avoid "a reduction in the number of media voices available to consumers."[18] Congress identified a specific interest in "ensuring [the] continuation" of "the local origination of [broadcast] programming,"[19] and found must carry necessary to serve certain broader aims of the Communications Act.[20] To promote localism and diversity in available programming, and to prevent cable operators from using gateway control over their distribution platform to exclude certain broadcast signals, Congress adopted the mandatory carriage provisions of the 1992 Act and later adopted somewhat different must carry requirements for satellite carriers. Since that time, the FCC has acknowledged the continuing importance of these rules and the growing incentive and ability of cable operators to use their powerful platforms to disfavor unaffiliated content.[21]

These requirements remain important to the nation's highly competitive system of free, over-the-air broadcasting today. Must carry stations tend to be stations that offer niche programming, such as foreign language, religious, or ethnic programming. These stations often target audiences whose needs are not being met by other programming sources. Because these stations serve narrower audiences, the ability to elect mandatory carriage is important to their continued survival. Carriage of these unique stations' signals is important to the diversity of both free over-the-air broadcasting and to the diversity of programming available via MVPD service.[22] I urge you to retain the current must carry requirements, rather than eliminating them as proposed in H.R.3675 and S.2008.

IV. Policies That Support Program Development, Innovation, and Localism Must Apply in the Same Manner to All Those Retransmitting Broadcast Signals

Some have raised questions about the role of Internet video providers in this regulatory landscape. Television broadcasters generally support the deployment of new and innovative Internet services, including broadband video services. Such services have the potential to enhance competition in the MVPD marketplace. Increased competition is a long-standing public policy goal, one that can be a positive development for consumers, broadcasters, and other program providers.

Greater platform choice, developed in a manner that respects the rights of content and signal providers, will provide benefits for consumers. For example, it is easy to see that consumers would benefit from the development and deployment of new, competitive distribution platforms capable of customizing programming or bundling different varieties of services, including voice, Internet access, and video services. Such customization may result in cost savings or increased access to programming of particular interest to the viewer.

Video programming providers, including broadcasters, may also benefit from the deployment of new video distribution platforms. The emergence of such additional platforms could provide programmers with additional outlets for reaching viewers and enhance video competition in the marketplace.[23]

Local television broadcasters, specifically, may also benefit from the emergence of new competitive MVPD services. New video distribution platforms represent other outlets for broadcast programming, including local news and information. These platforms could provide new opportunities for local broadcast stations to reach more local viewers and augment and enhance their program services to their communities. The advertising and retransmission consent revenues from these retransmissions would, in turn, be used to enhance news, entertainment, and public service programming – furthering the objective of localism. The emergence of another video distribution platform for carrying broadcast programming could also encourage greater innovation in digital television programming, including multicast and high definition ("HD") programming.

To achieve these public policy objectives, it is important that new services not be permitted to expropriate broadcast signals at will. Broadcasters must continue to have the right to control the distribution of their signals over the Internet and to obtain compensation from broadband video service providers seeking to retransmit such signals. If new technologies are allowed to evade retransmission consent and erode local viewership by overriding program exclusivity rights of local stations and offering the same programs on stations imported from distant markets, the viability of local TV stations – and their ability to serve their local communities with high quality programming – will be lost. For emerging video platforms that offer services comparable to those of MVPDs, the preservation of a fair, balanced, and symmetrical regulatory scheme would promote competition rather than impair it. No distribution platfrom – new or old – can under existing law (nor should it be permitted under any change in law) retransmit a station's broadcast signal without its consent. As Congress stated when it adopted the 1992 Cable Act, "broadcasters [must be allowed] to control the use of their signals *by anyone engaged in retransmission by whatever means*."[24]

CONCLUSION

It is an exciting time for consumers of video services. Viewers have more options for accessing content today than ever before. And the road ahead promises even better options as online video and other services bring innovation and competition to the marketplace. America's broadcasters are prepared to play a major role in the advancement of video services. Our historic transition to digital has vastly increased the amount and diversity of free, over-the-air television programming content available to nearly every home throughout the country and helped free up more than 100 MHz of spectrum that can be used for advanced wireless broadband services.

Yet, there are many challenges ahead. The FCC's planned incentive auction of broadcast spectrum could, if done correctly, ensure that local TV broadcasting remains the bedrock of our communication system. If done incorrectly, however, it could prove a disservice for millions of viewers that rely on over-the-air television as their primary information lifeline to the world. We respectfully encourage the Committee to keep a watchful eye on this process and act, if necessary, to ensure the FCC conducts the incentive auctions as Congress intended. We also encourage Congress to ignore the threadbare, self-serving arguments from a handful of pay TV companies for changes in the law for the sole purpose of securing a government-granted advantage in a highly competitive marketplace.

As the broadcast industry continues to develop new and exciting ways to deliver more diverse and high quality programming to the American people, my colleagues in the industry and I look forward to working cooperatively with the Members of this Committee.

Thank you for this opportunity to share our views with you.

End Notes

[1] *See* John Eggerton, "Study: Most Cord-Cutters May Be OTA _Opt-Ins,'" *Broadcasting & Cable*, June 18, 2012; *See also* blog post of David Tice, GfK Media Researcher, explaining

that what is commonly thought of as "cord-cutting" because of online video options may be better thought of as "cost-cutting," as consumers cancel expensive pay TV subscriptions and use free OTA television instead, available at http://www.gfkinsights4u.com/insights4u.cfm?articleID=511.

[2] Middle Class Tax Relief and Job Creation Act of 2012, Public Law 112-96, § 6403(a)(3).

[3] *See* Anton Troianovski, "Video Speed Traps Lurks in New iPad, Users Find the Superfast 4G Link Carries a Big Cost: Churning Through Data Limits in Mere Hours," *The Wall Street Journal*, March 22, 2012 (available at http://online.wsj.com/article/ SB10001424052702303812904577293882009811556.html?mod=WSJ_Tech_LEADTop).

[4] *See, e.g.,* Deborah D. McAdams, *McAdams On: Tangentially, Spectrum Policy Reform*, TVTechnology (June 8, 2012) available at http://www.tvtechnology.com/mcadams-on/0117/mcadams-on-tangentially-spectrum-policy-reform/213823 (discussing cable, satellite and telephone companies' "hedging" of prime spectrum, and quoting FCC Commissioner Robert McDowell as stating that the federal government occupies about 60 percent of the best spectrum and has no incentive to move off that space or to provide accurate information about the costs associated with moving); *See also* Randall Stephenson, *Spectrum and the Wireless Revolution*, Wall St. J., (June 10, 2012) available at http://online.wsj.com/article/SB10001424052702303665904577450222319683932.html (stating that much spectrum is held by "speculators" interested in making a profitable investment rather than building mobile networks, and arguing that such speculation should be discouraged and that regulations should be put in place to ensure that spectrum be used within a "reasonable" timeframe).

[5] Brian X. Chen, *Q.&A.: Martin Cooper, Father of the Cell Phone, on Spectrum Sharing*, New York Times Blog (May 31, 2012), available at http://bits.blogs.nytimes.com/2012/05/31/qa-marty-cooper-spectrum-sharing/.

[6] During the 2010-2011 television season, broadcast programming dominated the primetime program rankings, accounting for 95 of the top 100 programs. Source: The Nielsen Company, 9/20/10-5/25/11; Programming under 25 min. excluded; Ranked by AA% (ratings); in the event of a tie, impressions (000's) are used as a tiebreaker.

[7] *See* Declaration of Jeffrey A. Eisenach and Kevin W. Caves at 22 (May 27, 2011), attached to NAB Comments in MB Docket No. 10-71 (filed May 27, 2011).

[8] *See* Dave Seyler, *Broadcast Is Not Busting the MVPD Bank*, TV BUSINESS REPORT (June 8, 2012), available at: http://rbr.com/broadcast-is-not-busting-the-mpvd-bank/.

[9] As one party observed in comments filed with the FCC, "[c]able operators pay more than 10 times the per-subscriber fee for cable networks that are less than half as popular as the network-affiliated broadcast channels." Comments of the CBS Television Network Affiliates Association, MB Docket No. 10-71 at 14 (filed May 27, 2011).

[10] *See* Declaration of Jeffrey A. Eisenach and Kevin W. Caves at 30 (May 27, 2011), attached to NAB Comments in MB Docket No. 10-71 (filed May 27, 2011).

[11] *See,e.g., Letter from Steven Broeckaert, Media Bureau, to Jorge L. Bauermeister, Counsel for Choice Cable T.V.*, 22 FCC Rd 4933 (2007) (cable operator failed to meet good faith standard); *EchoStar Satellite Corp. v. Young Broadcasting, Inc.*, Memorandum Opinion and Order, 16 FCC Rcd 15070 (2001) (broadcaster met good faith standard while complaining MVPD was admonished for abuse of FCC processes and lack of candor).

[12] J. Eisenach and K. Caves, *The Effects of Regulation on Economies of Scale and Scope in TV Broadcasting*, at 3-4, attached to NAB Reply Comments in MB Docket No. 10-71 (filed June 27, 2011).

[13] The program exclusivity rules include the network nonduplication rules, *see* 47 C.F.R. §§ 76.92-76.95, 76.120-76.122, and the syndicated program exclusivity rules, *see* 47 C.F.R. §§ 76.101-76.110, 76.120, 76.123-76.125.

[14] The FCC's rules only (i) provide a forum for adjudication of program exclusivity disputes; (ii) limit and restrict the geographic scope of a program exclusivity arrangement between a

program supplier and a local television station; and (iii) impose certain formal notice requirements on local television stations as a condition to enforcement.

[15] *See, e.g., Implementation of the Cable Television Consumer Protection and Competition Act of 1992*, Memorandum Opinion and Order, 9 FCC Rcd 6723 (1994), at ¶ 114; S. Rep. No. 102-92 (1991), at 38.

[16] *Retransmission Consent and Exclusivity Rules: Report to Congress Pursuant to Section 208 of the Satellite Home Viewer Extension and Reauthorization Act of 2004* (Sept. 2005), at ¶ 50 (*FCC Report*).

[17] *See* 47 U.S.C. § 521 note (1992 Act § 2(a)(15)) (finding cable operators have an "economic incentive" to "refuse to carry new signals" from broadcasters, and that absent a must-carry requirement, "additional local broadcast signals will be deleted, repositioned, or not carried").

[18] *Id.* at 1992 Act §2(a)(4).

[19] *Id.* at 1992 Act §2(a)(10).

[20] *Id.* at 1992 Act §2(a)(9) (must carry is necessary to meet goal of "providing a fair, efficient, and equitable distribution of broadcast services") .

[21] *See Carriage of Digital Television Broadcast Signals*, Third Report and Order, 22 FCC Rcd 21064 (2007), at ¶¶ 49-52 (finding that because of increasing cable subscribership, rising audience shares for cable networks, cable's increasing share of the advertising market, and other factors, the cable industry was even stronger vis-à-vis broadcasters than in 1992 and that "cable operators have even greater incentives today to withhold carriage of broadcast stations.").

[22] *See* George S. Ford & John D. Jackson, *Preserving Free Television? Some Empirical Evidence on the Efficacy of Must-Carry*. JOURNAL OF MEDIA ECONOMICS, *13*(1), 1–14 (2000) (must carry helps preserve free over-the-air television, especially stations not affiliated with the four largest networks).

[23] In economic terms, the emergence of new outlets and distribution platforms will allow broadcasters, by disseminating programming to a wider audience, to take advantage of economies of scale and reduce their average cost per viewer. J. Eisenach and K. Caves, *The Effects of Regulation on Economies of Scale and Scope in TV Broadcasting*, at 6, attached to NAB Reply Comments in MB Docket No. 10-71 (filed June 27, 2011).

[24] S. Rep. No. 92-102, 1992 U.S.C.C.A.N. 1133, 1167 (1991) (emphasis added).

In: Future of Audio and Video Markets ... ISBN 978-1-62257-764-4
Editors: C. Powell Long, F. McDonald

Chapter 8

STATEMENT OF MICHAEL P. O'LEARY, SENIOR EXECUTIVE VICE PRESIDENT, THE MOTION PICTURE ASSOCIATION OF AMERICA. HEARING ON "THE FUTURE OF VIDEO"*

THE FILM AND TELEVISION INDUSTRY

Chairman Walden, Ranking Member Eshoo and members of the subcommittee I want to thank the Committee for the opportunity to testify on behalf of the Motion Picture Association of America, Inc.[1] and its member companies. As the primary voice and advocate for the American motion picture, home video and television industries in the U.S. and around the world, we are committed to providing audiences with a wealth of delivery options to experience the high quality entertainment that our companies produce. Whether you are at the theater, at home, on the road, or choose to download and view when you want, content creators, consumer electronics companies, and content distributors are working together to provide these new, innovative options for audiences to easily access licensed content. We welcome this opportunity to testify and work with the subcommittee as you consider appropriate policies. We hope you will recognize that as wondrous as these new forms of distribution are, they still need high quality content to deliver.

---

* This is an edited, reformatted and augmented version of the testimony given on June 27, 2012 before the House Committee on Energy and Commerce.

ENHANCED OFFERINGS TO OUR AUDIENCE

We have listened to consumers, we are attuned to their desires, and are at the forefront of developing innovative ways to meet and anticipate those desires. For some, there is not, and may never be, a better way to truly appreciate the magic of movies than in the theater setting. Creating wonderful movies that people can watch in movie theaters is an important part of America's rich history, but is equally part of the present and the future of video. In addition to utilizing cutting-edge and sophisticated visual and special effects in our productions, our companies have embraced 3-D, IMAX, and brilliant sound that has enhanced and even transformed the movie theater experience.

However, our member companies are not just about making movies for the big screen. We know that our audience enjoys watching our movies and television shows on every type of screen, from the television set in your living room, to your computer, iPad, Kindle, other tablet, or smartphone. We are partnering with companies of every stripe across the globe, including YouTube, Facebook, Netflix, Vudu, Roku – and soon, no doubt with some new platform being conceived by an unknown start-up company. All of these exciting innovations in distribution benefit both the audience who receive a high-quality viewing experience, and the creators who are able to monetize their enormous investment and success, and have the resources to reinvest in those dollars into creating even more high-quality programming.

VIEWERS CHOOSE SCREEN SIZE

For many people, watching great entertainment on their television with their family is a cherished part of the day. And today, our programs are being delivered to television screens by over-the-air digital broadcasts and through an astonishing range of channel choices enabled by our distribution partners in the cable and satellite business. Viewers are able to watch these programs at the time of original airing, or at a time of their choosing, either through DVR, or through the increasing array of video on demand options.

But a viewer's options do not end there. In addition, with the advent of interactive televisions, Blu-ray players, Roku boxes, Microsoft's X-box 360, Sony's Playstation, Nintendo's Wii, and Apple TV, viewers can watch HD movies and TV shows streamed across the Internet into their home through services like Netflix, VUDU, Hulu, HBO GO, or other services too numerous to be listed. And as Internet-enabled televisions continue to gain in popularity, the variety of options will expand even more.

More and more, people want the option to view programs not only at the theater or on TV, but also on their computers, tablets, and mobile phones. And we are responding to this call. Almost all the great ways you can watch your favorites programs on television are also available on-line or through your mobile phone. Hulu has been a great success, allowing people to use the Internet to catch up on their favorite TV shows or to explore new ones. YouTube and Facebook have entered into arrangements to distribute some of our companies' titles and Netflix is, for the first time, providing its customers with original content with shows like "Arrested Development". HBO GO is just one of the many services that allow cable and satellite subscribers to access the same content that they could on television and sometimes even more.

And in our enthusiasm over the distribution mechanisms, let's not forget Blu-ray disc, and the DVD, which much of our audience still consider the most enjoyable and convenient way to experience entertainment.

Our Audience Wants High Quality Content Online

It's amazing to think that in only a few short years the picture quality of online video has come to match traditional video distribution mechanisms. More than 350 unique licensed online services around the world now provide high-quality, on demand film and television shows online to viewers, including more than 60 such services in the United States. These services cater to every manner of consumer viewing model including rental viewing, download-to-own, subscription viewing, and ad-supported viewing. They are provided by every conceivable type of commercial entity, including technology companies like Apple's i-Tunes; broadcast television networks like ABC, CBS, Fox and NBC; cable networks like Comedy Central, TBS and USA; pay television channels like HBO and Epix; telecommunications/cable/ satellite providers like AT&T, Dish and Comcast; retailers and "rentailers" like Amazon, Best Buy, Blockbuster, Netflix, and Walmart (Vudu); gaming systems like PlayStation and Xbox; and new ventures devoted entirely to delivering great content seamlessly like Crackle, Hulu and MUBI.

Our Audience Wants Content Available on Their Mobile Devices

All of the major motion picture studios distribute full-length films and television shows directly to viewers' mobile devices through the major mobile operators. AT&T's MobileTV service contains full length episodes of hit television shows; television channels; and PIX, a film service for mobile phones launched by Sony. Sprint TV includes television channels and on demand television shows and movies from studios including Disney (Watch Disney Channel, Watch Disney XD, and Watch ESPN apps), Paramount, Sony, and Warner Bros. Verizon offers V Cast Mobile TV with channels and on demand full episodes of hit shows. Other wireless networks have similar licensed offerings. Apart from streaming to mobile devices, studios have made content available to load onto those devices through media such as SD flash memory cards, like those used in digital camera card.

Offering Portability and Flexibility to Our Viewers

More of our audiences want the convenience to access the content they purchased on a variety of devices without having to buy the same thing twice. And we're delivering. This began a few years ago, as our companies started to include with many DVD and Blu-ray disc titles a transferable or downloadable digital copy for viewers to use on computers or portable devices. Now there are a variety of ways to "buy once, and play everywhere". The Digital Entertainment Content Ecosystem (DECE) consortium of more than 60 studios, retail stores and technology firms has created "UltraViolet," a digital storage locker for your content. It works like this. When a consumer purchases UltraViolet media – such as a Blu-ray, DVD or Internet download – the viewer also receives the enduring right to access the content on any UltraViolet device registered to their household, and to enjoy the content via streaming through devices at home or on the go.

The first UltraViolet content rolled out late last year, through services like Walmart's Vudu, Flixster, and studio offerings. Presently, over 5,000 titles are available through UltraViolet and over three million consumers have set up accounts. This past spring, Walmart began offering the ability to convert current DVD and Blu-ray collections to digital UltraViolet copies that can both download and stream. Walmart estimates that the average user of this service is bringing about seven discs to the store and each conversion takes about a minute. Other [s] complementary digital initiatives are also being developed, such as Disney Studio All Access, which will provide viewers with easy access to Disney content across multiple digital video services and devices.

A Unique American Manufacturer

The U.S. motion picture and television industry is a creative community of many thousands of creators and makers, one that serves as a global cultural ambassador and one that plays a unique role in the American economy. Our members:

- Provide high-paying jobs to workers in all 50 states;
- Fuel small business growth, injecting capital into local, state, and national revenue pools and consistently generating a positive balance of trade;
- Help provide jobs for 2.2 million American workers who depend on the entertainment industry for their jobs, 12 percent of whom are directly employed in motion picture and television production and distribution – from behind-the-scenes production technicians to make-up artists and set-builders – across all 50 states;
- Provide quality jobs paying an average salary of nearly $79,000, 73 percent higher than the average salary nationwide;
- Support a nationwide network of nearly 95,000 small businesses—83 percent of whom employ fewer than 10 people—involved in the production and distribution of movies and television.
- Generate more than $15 billion in public revenue nationwide; and
- Return a positive balance of trade, making us critical to the U.S. export economy.

Over the last year our studios have filmed in Tennessee, Georgia, Illinois, Texas, New York, California, Michigan, Louisiana, Ohio and Massachusetts. Today there are ongoing productions in most of those states but we are also in Florida, Oregon and recently wrapped a film in Pennsylvania. "The Lone Ranger" is currently on location in Colorado and moves onto Utah next week.

On average, a major motion picture shooting on location contributes $225,000 every single day to the local economy, impacting the hundreds of thousands of other businesses that provide services to productions, like the local drycleaner that served the cast and crew on location or the local hardware store that supplied paint and lumber.

Audiences haven't been shy about their desire for seamless access to their favorite movies and television shows, at the time and on the device or platform that works best for them. We are relentlessly innovating to keep meeting that demand, and the choices available to audiences keep getting better. Our creative community – the studios and many thousands of creators and makers who create and make great movies and television shows – is committed to continuing that innovation. As noted in this testimony, we welcome and embrace the plethora of new forms of distribution, because they provide us new ways of offering our content to our audience in the manner they want to view our product. As this committee considers the appropriate policies for the future of video, we hope that it will recognize that as wondrous as these new forms of distribution are, they still need content to deliver. A video streaming site, a satellite television company, a broadcaster – they are all great technologies, but virtually meaningless to consumers if they don't have content to distribute. As the creators of that content, we are the entities that are taking financial risks in investing in product with no guarantee it will be profitable. As the party that takes the financial risk, we are entitled to control the distribution of our product. Make no mistake – we will always have an incentive to seek out new ways for our content to be distributed, but we must be allowed to be compensated for it and we must be allowed to make decisions about the proper context for its distribution. Mr. Chairman, Ranking Member Eshoo, again I thank you and this Committee on behalf of our member companies for the opportunity to testify today.

End Note

[1] The Motion Picture Association of America and its international counterpart, the Motion Picture Association (MPA) serve as the voice and advocate of the American motion picture, home video and television industries, domestically through the MPAA and internationally through the MPA. MPAA members are Paramount Pictures Corporation, Sony Pictures Entertainment Inc., Twentieth Century Fox Film Corporation, Universal City Studios LLC, Walt Disney Studios Motion Pictures, and Warner Bros. Entertainment Inc.

In: Future of Audio and Video Markets ... ISBN 978-1-62257-764-4
Editors: C. Powell Long, F. McDonald

Chapter 9

TESTIMONY OF MICHAEL K. POWELL, PRESIDENT AND CEO, NATIONAL CABLE AND TELECOMMUNICATIONS ASSOCIATION. HEARING ON "THE FUTURE OF VIDEO"*

Good morning Mr. Chairman and Members of the Subcommittee. My name is Michael Powell and I am the President and Chief Executive Officer of the National Cable & Telecommunications Association. Thank you for inviting me today to testify on the future of video. We welcome this important hearing.

CABLE ALWAYS HAS BEEN AN INNOVATIVE FORCE IN VIDEO

From its beginning, cable has driven innovation and transformation in the video business. Cable was founded to make broadcasting better -bringing it to suburban and rural areas outside the reach of over-the-air reception. We made programming better-breaking the lock of the three channel universe by investing billions in original content that appeals to specialized audiences as well as the mass market, and building award-winning iconic brands like CNN, ESPN, HBO, CNBC, C-SPAN, History and Discovery. We were first to

* This is an edited, reformatted and augmented version of the testimony given on June 27, 2012 before the House Committee on Energy and Commerce.

unshackle consumers from "appointment TV" with video on demand and the wide deployment of DVRs.

When we turned to areas other than video, the results were similar. In 1996 Congress wanted telephone competition and cable delivered it. Today, 1 in 3 households that have wireline phone service receive it from a cable operator.

And then there is broadband. Where high-speed data service was once the purview only of businesses, cable operators brought broadband Internet service to residential subscribers. This was not serendipity. The industry borrowed heavily and took enormous risk by ripping out its one-way analog network and replacing it with a higher capacity, two-way digital platform that made broadband possible. Cable broadband now reaches over 93 percent of homes-rich, poor, urban and rural. We have increased broadband speeds over 900 percent in a decade, (with even faster speeds coming this year) and we have extended the reach of our broadband service through extensive Wi-Fi networks. All of that in turn has enabled consumers to receive high-quality video over the Internet and on mobile devices.

The Golden Age of Video is Today

Television of the past was the rusty Iron Age compared to today. By any measure, the video experience has never been better than now. There is (1) more content, (2) higher quality programs, (3) more variety and diversity, (4) more sources for content-and video is available on a wider range of devices. There are hundreds of video programming networks, presented in brilliant HD quality and some in 3D. This is an enormous expansion from just twenty years ago. Artists and creators are producing some of the most compelling programs ever and cable is their preferred palette. In the most recent award seasons, cable programming won 73 percent of TV Golden Globes and 50 percent of Primetime Emmys. Acclaimed series such as "Game of Thrones," "Mad Men," and "Homeland" are some of the most accomplished dramas ever.

Public policy has always been concerned about diversity of viewpoints and niche programs for smaller yet passionate audiences. The cable model brought that ambition to fruition. The cable dial runs the gambit -from compelling scripted dramas, educational content, kids programming, sports, cooking shows, and news and public affairs. Simply put, if you fish, cook, workout, love music, crave sports, gorge on politics, admire dance, have a fetish for Jersey youth, or want programming in your native language or reflective of your community you will find it on cable. With all respect to former FCC Chairman Newt Minow, television today is no vast wasteland.

If you are itching to watch video, the number of sources you can turn to have grown exponentially as different providers compete for your business. You may subscribe to cable television and get up to 150 HD channels, the latest premium content and live events, video on demand and the ability to record and watch at your convenience on a DVR. You can get a very similar experience from Direct TV and Dish. Or, in many markets, you can also choose service from a telephone company. AT&T U-verse and Verizon FiOS have been popular new sources of video. Besides these subscription TV services, you can buy DVDs at Best Buy or better yet walk to the comer store and rent them for $1.20 from one of 30,000 plus Redbox kiosks.

If that were not exciting enough, Internet-delivered video has ushered in an even greater explosion of choice. Use of the Internet to watch video has increased 79.5 percent between the 3rd quarter of 2008 and the 3rd quarter of 2011.[1] By one estimate, real-time video streaming represents 62.5 percent of downstream Internet traffic in North America during prime time evening hours.[2] By the end of 2011, more than 100 million Americans watched online video content on an average day. That's a 43 percent increase from just a year ago.[3] Revenue from video content delivered over the Internet to televisions "is expected to grow from $2 billion in 2009 to over $17 billion in 2014."[4] The largest subscription video provider in the country today is Netflix -not Comcast, Time Warner Cable, DirecTV or any other multichannel video programming distributor (MVPD).

Companies that stream content are proliferating: Netflix, Vudu, Hulu, Amazon, iTunes, CinemaNow, Network websites, HBOGo, and user-generated or special interest sites like YouTube, Vimeo, and TED.com are a few. Many more offerings are anticipated from the likes of Google, Apple, Intel and Sony.

And web video is not limited to a PC screen any more. Computers can connect to big screen televisions; content can be beamed to sets using functions like Apple Airplay; box companies like Roku, Tivo and Boxee can deliver web video to the TV set; and manufacturers like Samsung are making the Flat Panel TV web enabled, with apps incorporated for accessing video content. One study estimates that at least 38 percent of U.S. households have a television set connected to the Internet, through either an Internet-ready TV, game console, standalone Blu-ray player or smart set-top box connected to their home network."[5] Smartphones and iPads have proliferated as compelling devices for consuming video content and enjoying second screen expenences.

Cable, too, is working to bring better video experiences to consumers wherever and whenever they want, offering, for example, applications that allow subscribers to view VOD and subscription linear channels on their iPads. Cable's "TV Everywhere" initiative makes it possible for our customers to watch video content they have already paid for on their laptops, tablets, smartphones and other portable devices-no matter where they are.

Even while offering this extraordinary array of new services, features and capabilities, cable has remained a solid value for American consumers. The fact that so many Americans stuck with cable during the recession is a testament to the value Americans get for their entertainment dollar. The price per viewing hour of entertainment for cable television is lower than for Netflix, a trip to the movie theatre, a sporting event, or even a DVD rental. Generally, the price for broadband service from most cable providers remained unchanged for 10-12 years following the introduction of services, even as broadband speeds increased nearly ten-fold over the same period of time.

Creators have never had so many ways to reach viewers. Broadband networks and the proliferation of devices capable of displaying video have dismantled the old barriers that separated content creators from their potential audience. A growing array of video content is being conceived and developed primarily for Web distribution. For example, Netflix has begun developing its own exclusive television series offerings to compete with video networks and services. Google is hosting original programs of all niches by independent producers. Of particular note, this popular new platform for accessing video content has emerged and thrived without any government-guaranteed access to programming or restrictions on licensing imposed upon content providers.

While not a complete substitute for a cable subscription, online video offerings are good enough for some consumers -currently a small minority -to cut or shave the cord. According to one report, in the three years between 2008 and 2011, "more than 2.65 million subscribers ... dropped their pay-TV service entirely in favor of streaming video options."[6] Yet during that same three-year period of time, 2.1 million net, new multichannel video customers were added in the U.S. While cord-cutting and cord-shaving are a business reality facing cable operators, it is also the case that there are millions of new customers subscribing to our service for the first time -or returning to us -because of the HD, on-demand, multi-screen and other advanced video offerings we make available.

As always, the cable industry is responding to changes in technology and in the marketplace in order to stay ahead of the curve and provide leading-edge services to its subscribers. The opportunities presented by broadband are great, as are the challenges. There are capacity constraints arising from the growing demands on the network imposed by services and offerings that require more throughputs. High-quality video depends upon continuous bit streams, but the Internet is largely a packet-switched network that sends traffic in bursts. Any network that is shared by many users-as all residential broadband networks are-must cope with congestion and ensure all users get quality service. And as we continue to invest billions annually in fixed cost networks, we must continue to explore fairer and more efficient ways to price service, including potential models that require heavier users to pay more for the additional network resources they use, rather than have lighter users subsidize the power users.

The Internet is vastly more flexible than traditional voice or video networks because it transmits information in packets and reassembles them at the subscriber device -but the "network of networks" that comprise the Internet does not yet offer the throughput and reliability to deliver video programming with the consistency of plant that is specifically built to deliver cable and satellite television services. By some estimates if you tried to put all cable content on the Internet, the Internet would be 100 times too small to handle it. While that circumstance will change, it will require investment and innovation to make that happen.

Cable's business incentives in today's marketplace are, contrary to suggestions from some quarters, fully aligned with the interests of consumers. The path to continued growth for cable is to enhance and expand its customers' use and enjoyment of the broadband platform we offer.

If consumers want to access video content via their laptop, their Xbox, their iPad, or their mobile device, it's our job to make that possible for them. If they want to obtain video content from Netflix, Amazon, Hulu, YouTube, Apple or any other online provider, it's our business to make that possible as well and we are. As our new tag line says, "Cable: It's more than TV. It's how we connect." Cable has a profitable broadband business and benefits from greater Internet consumption generated by streaming video services. To make the point, cable companies frequently promote web services like Netflix, Skype and Facebook in their ads.[7] So while it is natural that cable operators are themselves developing new services and features that enable their subscribers to access video online and on-the-go, such offerings do not in any way displace or hinder any other provider of content, service or devices in the online video market.

Claims and concerns that cable operators will use their control over broadband networks to stifle online video are simply not borne out by the facts. The exploding growth of online video usage undercuts any argument that cable is standing in the way of this business. To the contrary, cable's broadband networks are the engines that have enabled online video service to emerge and flourish. Indeed, there is simply no way to reconcile such claims with the wealth of content, services and devices that are available for accessing and enjoying online video.

THE FUTURE

As vibrant as today' s video marketplace is, the landscape of tomorrow that is taking shape is even more exciting. Content will be produced and distributed not just for television, but for any device with a screen capable of displaying video. Consumers will enjoy more variety and choice in how to access and interface with the video and on-demand offerings available to them from both their television and the Web, using, for example, their iPad or other portable device for navigation, instead of a remote control.

We can envision a world free of set -top boxes. The marketplace of the future is apt to move from being hardware-centric to software-centric, which will accelerate the pace of innovation and adaptation to new features and capabilities. Today, set-top box-based services cannot be deployed until the boxes are tested and glitch-free; by contrast, software-based services can be released in Beta and then improved. Beta versions of new products are common among the Googles and Apples of the world-and consumers benefit from the more rapid innovation under this model.

A vast array of cloud-based services and applications will make available a new generation of interactive and multimedia offerings, and dissolve the lines between video, data, graphics, voice, and text. Comcast is offering consumers a more personalized and interactive viewing experience with the launch of its Xl cloud-based DVR user interface. With the Xl, all the processing is done in the network, which gives Comcast the flexibility to quickly test and create new apps for customers without worrying about whether the customer's set-top box is outdated or not. It will also provide more personalized features, for example, allowing customers to see which shows their Facebook friends are watching. The Xl can also be used with mobile phone apps that turn handsets into remote controls.

Cable's investment in interoperable Wi-Fi networks will provide consumers with unprecedented levels of portability and flexibility for consuming video and other broadband service offerings. Cable Wi-Fi also enhances the efficiency of mobile broadband networks by enabling providers to offload traffic where Wi-Fi is available. Smartphones can use Wi-Fi to surf the web and even make calls, providing consumers with the opportunity to save money on their wireless service bills.

Importantly, all of the innovation that is taking place to bring both today's-and tomorrow's-advanced services is occurring *without* technology mandates not because of them and sometimes even in the face of existing regulatory impediments.

STRENGTHENING INCENTIVES TO INVEST AND INNOVATE BY UPDATING THE REGULATORY FRAMEWORK

The future looks bright, but it will require a business and regulatory framework that preserves and strengthens incentives to invest and innovate. The last 20 years have shown that cable operators and programmers are willing to invest aggressively to bring new services, features and capabilities to consumers. But legacy rules rooted in outdated assumptions concerning the state of competition in this marketplace will only frustrate our ability to obtain the capital for those investments. Investments are deterred by policies that seek to turp our networks into commodities. I doubt anyone here would want to revive the almost non-existent pace of innovation that residential telephone consumers experienced in the 50 years between 1940 and 1990, but that is the risk attendant to proposals that would turn network capacity into a commodity. The way forward to the future is not to revive regulatory anachronisms, but to cast off that baggage.

That means doing away with rules predicated upon marketplace conditions that no longer exist. Twenty years ago, cable was effectively the sole provider of multichannel video programming service in the country, serving 98 percent of all multichannel households. Today, incumbent cable's share of the multichannel marketplace stands at 56 percent and 4 of the 8 largest MVPDs in the country are non-cable.

Twenty years ago, there were no DBS companies providing multichannel video programming. Today, DirectTV and Dish Network are, respectively, the second and third largest MVPDs in the nation. Their multichannel video services are available to virtually every household in the country and together they serve 34 million households.

Twenty years ago, there were no telephone companies providing video service to cable subscribers. Today, AT&T and Verizon, the first and second largest communications companies in the world, make available multichannel video programming to more than 45 million households combined. Verizon and AT&T are respectively the sixth and eighth largest MVPDs in the country, and telephone companies are offering competing video service in rural as well as urban markets around the country.

Twenty years ago, the average cable system featured 36-54 channels and there were no more than 100 programming channels available for distribution to subscribers. Today, most consumers receive service from providers offering hundreds of channels, and there are now approximately 900 non-broadcast channels available for distribution to multichannel households.

Twenty years ago, vertical integration between cable systems and cable programming channels was widespread. Cable operators owned most national cable channels carried on a cable system. Today, vertical integration is the exception rather than the norm, falling from 53 percent of networks available for distribution in 1992 to only 14 percent today.

Twenty years ago, most American households relied primarily on over-the-air broadcast stations for their video entertainment. In fact, 41 percent of households exclusively relied upon broadcast television, a number that has fallen today to 14 percent. On average, broadcast stations accounted for 68 percent of television viewership in 1992. So, on a typical day, just over 2 out of 3 TV viewers were tuned into broadcast stations. Twenty years later, that number has flipped with more than 2 out of 3 television viewers tuning into non-broadcast channels during a given day.

While the premises for cable regulation have been shattered by competition and technological change, old rules based on aging snapshots of the video marketplace remain in effect. There are rate regulation rules designed to serve as a proxy for market-based pricing, even though most American households have a choice of at least three MVPDs and millions may opt to forego multichannel subscriptions altogether in favor of Internet-delivered video.

There are program access rules designed to nurture facilities-based competition to cable. But cable's main MVPD competitors-DirecTV, DISH Network, AT&T and Verizon-are all some of the largest communications services providers in the nation. These companies hardly need the government's help to remain viable competitors to cable. And, in contrast to 20 years ago, the vast majority of programming networks are not owned or controlled by cable operators, mitigating concerns that bottleneck control of programming could somehow stifle the emergence of competitors to cable.

There are also content carriage obligations, such as leased access, whose usefulness has been obviated by the Internet; and program carriage rules designed to protect programmers unaffiliated with cable, even though the vast majority of programming carried on cable systems and available to consumers already is unaffiliated with cable and it has never been easier for new video content to reach consumers. It is also time to re-examine those rules that affect the cable carriage of broadcast signals, such as must carry, retransmission consent, and non-duplication rights, that were adopted in a world in which technology and competitive market forces were vastly different than they are today.

In light of the fundamental changes that have occurred in the marketplace over the last 20 years, Congress should reassess the continued need for these restrictions. A new dialogue is essential because the continued application of outdated rules creates uncertainty and marketplace distortions, encourages unproductive attempts at regulatory arbitrage, and dampens incentives to invest.

Thank you again for the opportunity to appear today. I look forward to your questions- and to the continued opportunity to consider these critical policy issues.

End Notes

[1] *The Cross-Platform Report-Quarter 3*, 2011 US, Nielsen.

[2] *Global Internet Phenomena Spotlight 1H 2012 North America, Fixed Access*, Sandvine Incorporated, May 16, 2012.

[3] *US Digital Future in Focus*, COMSCORE, Feb. 2012.

[4] U.S. Dep't of Justice, Competitive Impact Statement, *United States v. Comcast Corp.*, No. 1:11-cv-00106 (D.D.C. Jan. 18, 2011), *available at* http://www.justice.gov/atr/cases/f266100/266158.htm.

[5] *38 percent OF U.S. HOUSEHOLDS NOW HAVE A TV CONNECTED TO THE INTERNET*, Leichtman Research Group, Inc. (April 9, 2012), available at http://www.leichtmanresearch.com/press/040912release.htrnl.

[6] Mike Snider and Roger Yu, *Flood of Video Streaming Options Could Confound TV Watchers*, USA Today (April 10, 2012), *available at* http://www.usatoday.com/tech/news/story/2012-04-09/strcaming-vidco-options/54136024/1. See also, Daniel Frankel, Nielsen: 1.5M U.S. households cut the cord in 2011, PAID CONTENT, May 4, 2012, at http://paidcontent.org/2012/05/04/nielsen-1-5-m~u-shouseholds-cuH.he-cord-in-2011/.

[7] *See* http://www.youtube.com/watch?v=zEhuULSD3MAI.

In: Future of Audio and Video Markets ... ISBN 978-1-62257-764-4
Editors: C. Powell Long, F. McDonald

Chapter 10

BACKGROUND MEMO TO THE HEARING ON "THE FUTURE OF AUDIO"*

The Committee on Energy and Commerce

To: Members and Staff, Subcommittee on Communications and Technology

From: Majority Committee Staff

Re: Hearing on "The Future of Audio"

The Subcommittee on Communications and Technology will hold a hearing Wednesday, June 6, 2012, at 10:15 a.m. in 2123 Rayburn House Office Building on "The Future of Audio." One panel of witnesses will testify.

I. WITNESSES

Mr. Ben Allison
Bassist/Composer and Governor
New York Chapter, National Academy of Recording Arts and Sciences

* This is an edited, reformatted and augmented version of the testimony given on June 6, 2012 before the House Committee on Energy and Commerce.

Mr. David M. Israelite
President and CEO
National Music Publishers' Association

Mr. Cary Sherman
Chairman and CEO
Recording Industry Association of America

Mr. Jeff Smulyan
Chairman, President, and CEO
Emmis Communications

Mr. Steven W. Newberry
President and CEO
Commonwealth Broadcasting Corp.

Mr. Tim Westergren
Chief Strategy Officer and Founder
Pandora

Mr. Christopher Guttman-McCabe
Vice President, Regulatory Affairs
CTIA Majority

Mr. Gary Shapiro
President and CEO
Consumer Electronics Association

This is not our parents' audio market. In the last 20 years, broadcast radio has gone HD, bringing HD radio owners the advantages of digital transmission and as many as four channels each from more than 2,000 participating local stations. Satellite radio now offers subscribers more than 150 digital channels from anywhere in the nation. Portable MP3 players allow audiophiles to carry their entire music collections to the gym. And the Internet gives listeners access to a virtually infinite amount of legal (and pirated) content.

The number of outlets has increased exponentially. The Internet, in particular, has all but obliterated technological and economic barriers to entry into the marketplace. Anyone with access to an Internet connection who can speak, sing, or play an instrument—no matter how well—can reach the world's listeners for a song. But if a chord rings in the digital woods and no one hears it, does it make a sound? Paying the rent while launching your career may be a little easier now, but the growth in available channels has fractured audiences. In the past, getting on the radio brought a certain likelihood you would obtain at least a modicum of success (and remuneration). Today, everyone can be "on the radio." Getting from the garage to the Grammys may ironically be harder, requiring not only more talent, but also more marketing to rise above the din for anyone except those with the most viral of Internet videos.

These developments are stirring tensions among the participants in the distribution chain. Some fear they will garner a smaller share of the pie. Others fear they will be disintermediated altogether. The changes are also placing strains on existing legal frameworks. And somewhere amidst it all sits the artist. This hearing will examine how advances in communications services and consumer electronics equipment are affecting the distribution and consumption of audio content. What follows is background on some of the regulatory issues that may arise.

II. BACKGROUND

Broadcast Radio Ownership Caps

In light of the robust competition in the audio market, some broadcasters argue that certain Federal Communications Commission (FCC) restrictions on broadcast radio ownership no longer make sense. If the agency found enough competition to allow XM and Sirius—the only two domestic satellite radio companies—to merge in 2008, then there must be enough competition to liberalize the radio ownership limits, they say.

The newspaper-broadcast cross-ownership ban, put in place in 1975, prohibits an entity from owning in the same market both a newspaper and a broadcast radio or broadcast television station. Pursuant to the agency's obligation to review its broadcast ownership restrictions in 2006 as part of the statutorily mandated *Quadrennial Review*, the FCC sought to relax the ban in December 2007 in favor of a case-by-case review. Cross-ownership would be presumed permissible in the 20 largest markets and presumed impermissible in smaller markets. The U.S. Court of Appeals for the Third Circuit overturned the decision in its July 2011 *Prometheus Radio Project v. FCC* ruling, however, on the grounds that the editorial then-Chairman Martin used to propose the revision did not constitute sufficient notice under the Administrative Procedure Act. Opponents of the ban are once again urging the Commission to eliminate it in light of increased competition and because newspapers are struggling in the current marketplace.

The radio-television cross-ownership rule, put in place in 1999, allows an entity to own up to two television stations and four radio stations in a market, so long as at least 10 independent media voices would remain in the market after a combination. It allows an entity to own up to two television stations and six radio stations, or one television station and seven radio stations, in a market so long as at least 20 independent media voices would remain in the market after a combination. For purposes of the rule, independent voices are full-power broadcast TV stations, primary broadcast radio stations, daily newspapers, and cable systems. The FCC retained the radio-television cross-ownership rule in its 2007 *Quadrennial Review* order, a decision the Third Circuit upheld in its 2011 *Prometheus* decision. Opponents of the rule are again arguing that increased competition makes the rule unreasonable. They also criticize as irrational the omission of satellite and Internet services from the "independent voices" analysis.

The local broadcast radio caps, put in place in 1996, prohibit an entity from owning:

- more than eight commercial radio stations in markets with 45 or more stations, and no more than five can be in the same service (AM or FM);
- more than seven commercial radio stations in markets with 30-44 stations, and no more than four can be in the same service;
- more than six commercial radio stations in markets with 15-29 stations, and no more than four can be in the same service; and,
- more than five or more than 50 percent of the commercial radio stations, whichever is
- less, in markets with 14 or fewer radio stations, and no more than three can be in the same service, except that an entity may always own a single AM and single FM station combination.

The FCC retained the local radio ownership caps in its 2007 *Quadrennial Review* order. The Third Circuit upheld the decision in its 2011 *Prometheus* ruling, concluding that the agency's decision was not so arbitrary and capricious as to violate the Administrative Procedure Act. Opponents of the caps argue that doesn't mean they are good policy. They say that ownership of multiple radio stations in a market promotes diversity because station owners do not want to duplicate formats and compete with themselves. They also point out that the market has evolved even more since 2007. Between 1996 and now, the number of full-power radio stations has increased by 23.7 percent, satellite radio has gained more than 20 million subscribers, Internet radio garners more than 89 million listeners each month, and a majority of Americans over the age of 12 possess a portable music player.

As part of the 2010 *Quadrennial Review*, the FCC proposed in December 2011 to reinstate Chairman Martin's relaxation of the newspaper-broadcast cross-ownership ban and eliminate the radio-television cross-ownership rule, but to retain the local broadcast radio ownership caps.

Low-Power FM

In January 2011, Congress passed the Local Community Radio Act, introduced by Reps. Doyle and Terry, to allow authorization of additional low-power FM stations. In particular, the legislation removes the third-adjacent channel minimum distance separation requirements and allows waiver of the second-adjacent channel minimum distance separation requirements in certain circumstances. The legislation also makes clear that "FM translator stations, FM booster stations, and low-power FM stations remain equal in status" and imposes certain interference protections. The FCC issued an order and notice of proposed rulemaking in March 2012 to implement the law and seek further input.

FM Chip

A number of broadcasters seek a mandate requiring consumer electronics manufacturers and wireless providers to include an FM receiver in cell phones. They argue that requiring an FM chip in cell phones is necessary to preserve free, over-the-air radio. They also argue that the mandate is justified because over-the-air radio provides critical information in times of emergencies. Manufacturers and wireless providers oppose the mandate on the grounds that it will increase costs and that requiring an FM chip in wireless devices would duplicate the emergency alert program the wireless industry is deploying as a result of the WARN Act. They also argue that features of phones should be determined by consumer demand and say that more than 40 wireless device models already have an FM chip.

Royalties

Under copyright law, satellite and Internet radio stations must pay performers and recording owners for music they transmit. Broadcast radio stations do not, however. They argue this is necessary to preserve free, over-the-air radio, and that broadcast radio airtime promotes record sales. Performers and record labels, however, argue that they should be the ones to decide how much to value the promotional benefits of broadcast and should be compensated for use of their music. Satellite and Internet radio operators also say that it is unfair for their broadcast competitors to pay less than they do.

In: Future of Audio and Video Markets ... ISBN 978-1-62257-764-4
Editors: C. Powell Long, F. McDonald

Chapter 11

TESTIMONY OF BEN ALLISON, MUSICIAN AND GOVERNOR, NEW YORK CHAPTER OF THE NATIONAL ACADEMY OF RECORDING ARTS AND SCIENCES. HEARING ON "THE FUTURE OF AUDIO"*

Good morning Chairman Walden, Ranking Member Eshoo and members of the Subcommittee. My name is Ben Allison, I am a musician, composer and producer based in New York. I am also a board member of the New York Chapter of The Recording Academy. Outside the beltway, The Recording Academy is best known for producing the GRAMMY Awards. But here in DC, the Academy is the trade association representing music creators — more than 20,000 performers, songwriters and studio professionals. I want to thank the members for the privilege of being able to share my perspective as an artist, and to represent The Academy.

In less than 24 hours, I'll be traveling the Bonnoroo Music Festival in Tennessee to perform with my band. But the music created at Bonnoroo will not be confined to the 80,000 people attending the concerts. Like many festivals, Bonnoroo will offer live streaming of performances, while internet radio services like Pandora will offer Bonnoroo-themed stations. This is just one example of why I'm excited about the Future of Audio. New technologies will allow us to connect to more fans in new and innovative ways.

* This is an edited, reformatted and augmented version of the testimony given on June 6, 2012 before the House Committee on Energy and Commerce.

Internet and satellite radio are growing exponentially as a source for fans to access music. I'd like to focus on three issues regarding these new technologies: discovery, audio quality and compensation.

Discovery

As for discovery, the infinite possibilities of stations on Internet radio and the niche genres available on satellite have given more artists a chance to be heard that would never be played on AM or FM radio. The beauty of Internet radio services like Pandora is that song selection is blind to popularity and is simply based on matching tracks to the consumers' tastes. So artists without large promotional budgets will be played if the music suits the listener's ear, and a new fan may be born.

As a board Member of the Recording Academy's New York Chapter and a working musician on the New York music scene, I meet many young and emerging artists who are using these new services to connect to fans. And a number of niche styles of music that never get played on AM or FM radio are finding homes on Internet radio. And these platforms don't only provide promotion, they provide compensation—to artists and songwriters.

Audio Quality

Early on, the transition to a digital marketplace had led to the use of inferior listening formats and resulted in there being less of an emphasis on sound quality. One of the results of this is a "disposable" attitude toward consuming music—there is less perceived value in the music itself. When people actually hear higher quality sound the difference can be very apparent—like going from black and white to color.

As an acoustic musician, the quality of the sound delivered is of the utmost importance to me. I've chosen my instrument, an 1840 American-made bass, for its pure and nuanced sound. So I want my listeners to hear a recording that sounds as close to the original performance as possible.

The Recording Academy's Producers and Engineer Wing has been very active on the issue of sound quality, working with labels, technology companies and consumers to raise the level of quality.

And now, consumers are waking up. There is a renewed interest in higher quality audio, as evidenced by the desire for premium headphones and higher resolution downloads. Increased bandwidth will make it easier for consumers to access higher quality audio and to make listening to it commonplace again.

Compensation

Noninteractive Internet and Satellite radio services primarily pay performers through a compulsory license, allowing the services to play any song they choose, as long as they pay the rate established by the Copyright Royalty Board set up by Congress. The payments are processed by SoundExchange, a nonprofit entity with oversight that includes artist representatives. SoundExchange ensures that 50% of all payments go directly to performers, and in the past five years, payments to artists through SoundExchange have increased by 400%. The ease of payment to one entity has allowed new businesses to emerge without the burden of multiple licensing deals. And because of this ease of payment to SoundExchange, services have almost exclusively taken advantage of the compulsory license. However, one concern for artists is that services may in the future choose to license directly with record labels. Such a scenario could mean lower payments to artists, which is why performers generally prefer their payment be made through SoundExchange.

Another concern has to do with interactive, or "on demand" services. Unlike Internet radio, the consumer can choose to listen to a particular track whenever they want. Streaming services such as Spotify are currently paying a very low royalty rate—a fraction of a penny per spin. These arrangements are negotiated in the marketplace by the services and copyright owners. As artists, we are told the rates will grow. I hope that's true. We're also told that we are benefitting from the promotion, but promotion should never replace compensation. If on-demand services replace downloads, then that reduces the compensation for creators—and this is a great concern.

The Future of Audio—and the Past

But, ironically, one of the most pressing issues for performers is not about the *future* of audio, but its past. And the concern is not caused by a *new* technology—but rather by one that has been around for more than a century: traditional, over-the-air radio.

Terrestrial broadcasters have an inexplicable "free ride" when it comes to performance royalties. They are exempt from paying performers any royalties when they use our recordings to fuel their multi-billion dollar industry. This makes corporate radio the only business in America that can legally use another's intellectual property without permission or compensation.

All of the other broadcast platforms—Internet, satellite and cable—pay a performance royalty for sound recordings, regardless of promotion. Every other country in the developed world has such a right. But in the U.S., over-the-air broadcasters are required to pay songwriters—as they should—but not performers.

The previous Congress made great progress in addressing this issue, which led to private negotiations. And in fact, after a year of negotiations, the National Association of Broadcasters and the musicFIRST Coalition (representing artists and labels) agreed on a settlement that would have created a terrestrial performance royalty in exchange for specific rate reductions on simulcasting. Unfortunately, when the NAB Board voted on the deal, they changed the agreed-upon numbers, making it a net loss for the music side, and the deal collapsed.

To add insult to injury, terrestrial broadcasters are now asking for another legislative perk—a mandate to include FM receivers in smart phones. They already exploit one anomaly in the law—they do not pay performers while new audio technologies do. Now they want a second legislative anomaly—to have a free ride—quite literally—on the backs of new devices.

Artists believe Congress should address the first anomaly before there can be any discussion about the second.

One last word on this point—just yesterday Clear Channel announced a deal to pay performance royalties to one record label. With the largest radio broadcaster agreeing that a terrestrial performance right should exist, there is no longer any legitimate argument for the NAB to oppose the right. Congress should resolve this issue once and for all and remove the corporate radio loophole.

Democratization of Music

Finally, I'd like to close by discussing the lower barriers to entry for music creators, often referred to as "the democratization of music." With new platforms for audio allowing nearly every musician to be in the virtual record store, there's an ongoing debate about how much music is too much for the consumer to absorb.

I'm a professional musician. I started taking music lessons at the age of nine, went to a performing arts high school, and graduated from NYU with a degree in music performance. I played my first gig at 16. This is my life and my livelihood.

That said, I am delighted that those for whom music is a part-time pursuit can have access to the digital marketplace. And some of those part-timers may be great musicians who might just find a fan base on Pandora or Sirius Satellite radio and be able to become full-time musicians. More music is always a good thing and quality will always find its place in the market.

But while it's fine for the market to welcome non-professional musicians, we must not allow the market to make music a non-profession. Recently, John McCrea, the lead singer of the band Cake, stated, "Can you put food on the table with music? Probably not. I see everybody I know, some of them really important artists, studying how to do other jobs."

Each one of my fellow panelists here has a role in connecting the music creator to the fan. But they'll have no business if there's no great music. Without the songwriter putting that first note on paper…without the musicians performing that song in ways that move us…without the producer and engineer capturing that performance…there would be no iPods, no Pandoras, no labels, no publishers.

We do not want the next generation of great artists studying how to do other jobs. Music must be respected, content protected online, and all creators compensated.

So let me close by stating one constant in the music industry. In this regard, the future of audio is the same as the past: it is dependent on the creator. And we, distinguished members of the subcommittee, are dependent on you to protect our rights.

Thank you.

In: Future of Audio and Video Markets ... ISBN 978-1-62257-764-4
Editors: C. Powell Long, F. McDonald

Chapter 12

TESTIMONY OF DAVID M. ISRAELITE, PRESIDENT AND CEO, NATIONAL MUSIC PUBLISHERS' ASSOCIATION. HEARING ON "THE FUTURE OF AUDIO"*

Good morning Chairman Walden, Ranking Member Eshoo, and Members of the Subcommittee. Thank you for the opportunity to appear before you today to talk about the future of audio as it pertains to music publishers and songwriters.

I serve as President and CEO of the National Music Publishers' Association ("NMPA"), the principal trade association representing music publishers and songwriters in the United States. Now in our 95th year, the NMPA represents over 2,800 member companies with the goal of protecting and advancing their property rights on the legislative, litigation, and regulatory fronts. Prior to this position, I also had the honor of serving at the Department of Justice where I chaired the Department's Task Force on Intellectual Property.

THE ROLE OF MUSIC PUBLISHERS

Issues involving the music industry are complicated, in part because there are two separate and distinct copyrights involved in music.

---

* This is an edited, reformatted and augmented version of the testimony given on June 6, 2012 before the House Committee on Energy and Commerce.

The first copyright is for the underlying musical composition created by one or more songwriters, and often owned or represented by a music publisher. I am here representing that half of the music industry. The second copyright is for any recording of that song – commonly known as the sound recording copyright – and represented by record labels. It is crucial to understand that these two different copyrights are controlled and represented by different interests, and are often treated very differently under the law and in business practices.

If that were not complicated enough, the different uses of the musical composition copyright are also treated differently under the law and through government regulation.

Songwriters and music publishers attempt to earn a living through three primary means of utilizing their separate copyright – mechanical reproductions, public performances, and audio-visual synchronizations. Each type of income represents roughly a third of their revenue.

First, there is the mechanical reproduction right. An example is when a consumer downloads a song from iTunes or streams music through a service like Spotify. For songwriters, this right is regulated by Section 115 of the Copyright Act which imposes a compulsory license system on the songwriter/publisher community. Songwriters and music publishers do not get to negotiate the value of our intellectual property in a free market. For record labels, this is a free market right and not regulated by law. The compulsory license dates back to 1909 – before the existence of recorded music – when Congress decided to regulate the mechanical reproduction of musical compositions embodied on piano rolls for player pianos.

Second, there is a public performance right. An example is when music is performed on the radio. While this right is inherently unregulated by law, the vast majority of the market is regulated by consent decrees with the Department of Justice. Again, under these consent decrees, songwriters and music publishers do not generally get to negotiate the value of their intellectual property in a free market. In addition, there are situations where the rights of songwriters and music publishers are utilized using the consent decrees, but there is no compensation for vast lengths of time due to provisions of the consent decree allowing for temporary free-rate licensing.

Third, there is the use of music synchronized with video. Traditionally this has included using music in movies, television, and commercials. For songwriters and music publishers, this is a free market right not regulated by law. Newer forms of this right include music videos and the use of music in user-generated content such as in YouTube.

THE FUTURE OF AUDIO

For songwriters and music publishers, the future of audio presents both opportunities and challenges. New technologies allow us to deliver music through innovative legal channels that give fans what they want. For example, the iPhone -- introduced just five years ago -- has revolutionized how consumers access and listen to music. And services such as Spotify -- just recently introduced in the U.S. -- give music fans unprecedented access to large music catalogues. We have embraced all of these developments and are looking forward to the next new services that will enhance the culture of music. However, there are also three challenges music publishers and songwriters face as we move toward the future.

First, we must do a better job of protecting music from theft. While there can be legitimate debate on how best to deal with new technologies, it can never be right to steal the intellectual property of songwriters. It also cannot be right for parties to facilitate or turn a blind eye to such theft while profiting from that theft. Second, we must find efficient ways to license our copyrights and empower new business models. Much of the current licensing system is outdated and inefficient. It was built to service outdated business models.

Finally, we must ensure that future business models fairly compensate songwriters. Licensing new business models efficiently does no good if such new business models do not allow a songwriter to earn a living.

THEFT

The impact of theft on the music world is a familiar subject to all of you. It is important to consider that it is significantly more difficult for songwriters to police the internet for infringement than it is for large corporations. Most songwriters simply do not have the resources or capacity to find the ever-increasing number of unauthorized uses of their works on the Internet, and to engage in the process of sending numerous take-down notices for each new unauthorized posting. Take-down notices to infringing websites now number in the thousands and even millions for some internet service providers. Many corporations are barely able to keep up. Songwriters have, to a large extent, simply stopped trying.

Songwriters are particularly reliant on receiving fair compensation for the use of their works online and through digital music services because unlike recording artists, many songwriters do not have the ability to generate income in other ways, such as through touring or merchandise sales. Alternative income streams are simply not available for most songwriters, which makes the theft of their music that much more devastating.

Licensing

Just a few months ago, 25 parties completed a year-long negotiation over rates for five new categories of music services to allow flexibility in creating new services that enable consumers to access, use and purchase music in previously impossible ways. These new categories cover things like mixed service bundles that bundle music products with non-music products, and cloud based locker services that allow consumers to store and access their own music across almost every electronic device. And, parties representing digital services, record labels, and songwriters and publishers are currently involved in discussions on how to work together to improve our licensing system.

Compensation

Right now, interactive streaming music services comprise a fairly small percentage of the income that songwriters collect. However, I believe the streaming market is one of the significant growth areas for music in the future. As this area of the market grows, we need to ensure that songwriters are fairly compensated for their work. It is also crucial that new services like Vevo properly license and compensate songwriters.

Conclusion

The marketplace for music is constantly evolving in ways that are both beneficial and potentially detrimental for music's creators. Songwriters and music publishers will continue to embrace new delivery models and technology, but as the future of audio develops Congress and the music community must ensure that laws protect intellectual property while providing fans the music they want in the manner they want.

In: Future of Audio and Video Markets ... ISBN 978-1-62257-764-4
Editors: C. Powell Long, F. McDonald

Chapter 13

STATEMENT OF CARY SHERMAN, CHAIRMAN AND CEO, RECORDING INDUSTRY ASSOCIATION OF AMERICA. HEARING ON "THE FUTURE OF AUDIO"*

Good morning Chairman Walden, Ranking Member Eshoo, and Members of the Subcommittee. Thank you for inviting me to testify this morning on "The Future of Audio." My name is Cary Sherman and I am Chairman and CEO of the Recording Industry Association of America. We represent America's major music labels.

I'm glad to be here today, because I believe that the recording industry, and the music industry in general, has a great story to tell. The bottom line is that the music industry today has transformed how it does business, and we expect the industry to continue to evolve, enabling new artists to prosper and allowing consumers to enjoy their works in many different ways.

Today, the music business earns more than half of its revenues from an array of digital formats. That's right, CDs are no longer the primary format for the music business or the primary way the industry generates revenues. Digital is not just our future, it is our present.

In 2004, the first year we had any meaningful digital revenues, the industry earned a grand total of $190 million from digital services. Last year, we hit nearly $3.5 billion. Quite a change.

* This is an edited, reformatted and augmented version of the testimony given on June 6, 2012 before the House Committee on Energy and Commerce.

We often hear the complaint that we need to get a new business model. Well…our companies have done just that:

- You want DRM-free downloads? We've got that: iTunes, AmazonMP3, eMusic, 7digital.
- You want to pay a modest monthly fee for all the music you can ever listen to – on your computer or smart phone? We've got that: Rhapsody, Spotify, MOG, Rdio, Music Unlimited, rara.com, Zune Music Pass.
- You want free, ad-supported video and audio streaming? We've got that: Spotify, YouTube, Vevo, Myspace Music, AOLMusic.
- You want music bundled with your mobile phone? We've got that: Muve Music, Metro PCS/Rhapsody
- You want to store all your music in the cloud, so you can access it from wherever you might be? We've got that: iTunes Match, and more deals in the works.
- You want specialized digital radio services that offer you the niche kind of music you like to hear? We've got that: Pandora, SiriusXM, Last.fm, Yahoo!Music, AOLMusic, and over750 more such services.
- You want online simulcasts of AM/FM radio stations? That is available too: iHeartRadio, WJLK-FM 994.3 (The Point), KPWR-FM (Power 106), WXLC-FM (102.3 XLC), and over 750 more online radio stations fully licensed through our sister organization SoundExchange.

There are over 500 digital services authorized by our member companies worldwide offering 20 million authorized tracks. ALL of these business models have been embraced and authorized by major and independent music companies.

And we're not stopping there. One of our highest priorities at RIAA is to develop the infrastructure that will make it faster and easier for entrepreneurs to offer even more innovative business models to music fans.

Just last month, we announced, along with my colleague David Israelite of NMPA and digital music services, a groundbreaking licensing agreement that will make it easier for digital services to clear publishing rights for five categories of new business models.

We're also working on new industry-wide databases and royalty distribution systems to make royalty payment functions more efficient; and licensing reform to update the statutory mechanism for the old "mechanical" licensing system. We are intent on working with our Internet and publishing partners to simplify and expedite the licensing process.

It probably goes without saying that we are seeking out and embracing these new structures and offerings under the continuing threat of rampant piracy. In fact, we're less than half the size we used to be: down to $7 billion in 2011 from nearly $15 billion in 1999. Certainly, piracy does not account for that entire loss. But nearly every academic study, and nearly every economist – not to mention common sense – has concluded that illegal downloading has hurt us badly. What kind of harm? Massive layoffs, of course. But also less money to invest in artists. That means fewer artists on our rosters, fewer people who can make a living from music, fewer songs permeating through our culture that help form a piece of our national identity. In fact, according to Bureau of Labor Statistics data from the Federal government, the number of people who identify themselves as "musicians" has declined over the last decade, conspicuously tracking the decline of the industry. Piracy is not just a parochial corporate problem. This is an issue that affects many industries, our economy, our culture, tens of thousands of creative individuals, and most importantly, the consumers who enjoy the music we create.

When it comes to protecting or enforcing creative rights, the effort is often caricatured as a quixotic game of whack-a-mole that only enriches the lawyers. But fresh evidence is emerging that strategic copyright protection combined with robust legal digital offerings *can* put money into the pockets of artists and songwriters and the companies who invest in them.

For example, most people are familiar with Limewire, which was the world's most popular peer-to-peer (P2P) file-sharing service. Between 2007 and 2010, about 2/3 of file-sharers on the Internet used the service. At the end of October 2010, a federal court finally shut them down for inducing massive copyright infringement. The very next month, digital sales of music improved and they've remained higher ever since. While this may not be the sole cause, it is not a coincidence. Just a couple of months ago, market research firm NPD released a survey showing that more than half of the people who had used Limewire did NOT go to another illegal music site to get their music after the shutdown, thanks to the availability of all those authorized services I mentioned.

Government enforcement of criminal copyright laws to protect our nation's economic interests is also vitally important. The indictment of Megaupload has had a tremendous impact on other such rogue cyberlocker sites. The government's action sends a signal that the United States will not tolerate the use of the Internet for criminal activity that violates our laws.

We continue to believe that the best and single most important anti-piracy strategy remains innovation — experimenting and working with our technology and Internet partners on consumer-friendly new business models. But enforcing our constitutional property rights is also a necessary part of the equation.

So how are we approaching protecting our rights these days? For the most part, by forging voluntary, marketplace agreements with others in the Internet ecosystem under which everyone plays a part in addressing the problem. Just last year, we announced a voluntary program with ISPs that will be implemented later this year to address illegal downloads on P2P networks. We also helped craft an agreement with major credit card companies and payment processors on voluntary best practices to reduce sales of counterfeit and pirated goods. And just last month, major advertisers and ad agencies announced a series of voluntary best practices so that their valuable brands are not associated with rogue Internet sites that offer illegal goods, and advertisers don't inadvertently enrich rogue website operators. We hope other intermediaries like search engines will follow suit in negotiating voluntary marketplace best practices to prevent directing users to sites that are dedicated to violating property rights.

These voluntary programs are not a panacea. No program ever will be. And sometimes, the Congress must step in to assure that our property rights, and U.S. economic interests, are being protected. Especially against sites overseas whose business model is the theft of U.S. works. But collectively, we think these collaborative efforts will make a difference. They are the product of outreach, and a lot of conversation over several years – not only with these intermediaries, but also with public interest groups who want to figure out how to address online problems while ensuring the reasonable preservation of a free and open Internet.

We need to engage in the same sort of outreach directly with the tech and Internet communities, and I am committed to doing that – because, in the end, we all have an interest in an Internet that is open and accessible, but not lawless.

Speaking of working together, I would be remiss if I didn't take this opportunity to once again point out a glaring inequity when it comes to compensating creators. The bottom line is that every platform that (legally) plays music pays to do so – except for one. AM/FM radio stations use music to draw billions of dollars in advertising revenue for themselves, but they don't pay a cent to artists, musicians and sound recording owners who make the music they use. Internet radio, cable music channels, and satellite radio all compensate artists and labels for the music they play, while promoting artists at the same time. All other copyrighted works, without exception, receive payment for their use, regardless of promotion. Radio stations in every other developed country in the world compensate artists and labels for the use of music, regardless of promotion. This extraordinary government subsidy for U.S. broadcasters at the expense of U.S. music creators is due to an unfair and unfortunate anomaly in our law that has persisted for decades, and must be addressed.

I do think many of us in the music industry find it frustrating that, instead of working with us to find a fair resolution to this issue, broadcasters are more focused on getting the government to mandate the insertion of an FM chip in mobile devices that was rejected by an intergovernmental advisory group. We suspect that the broadcasters' real agenda is to get an FM chip installed or activated in every U.S. cell phone, regardless of consumer demand, to prevent being overtaken by the popularity of Internet radio services on those devices, such as the one by our partners at Pandora. Pandora and others are making great strides with consumers on these devices, and challenging traditional broadcasters in the home, in the car and on the move. They are giving consumers what they want, based on consumer demand, and paying artists and record companies while they do it. Certainly, broadcasters like ClearChannel also offer Internet radio services like iHeartRadio that are available on cell phones and that go head to head against other Internet radio services on all mobile devices. Those services do have to pay artists and labels. But it is the old line, old-fashioned terrestrial service, the one that is exempted and subsidized, the one that does not pay artists and labels for the use of music, that broadcasters want to require on all mobile devices. That does not sound like moving forward to me.

These issues are important. But in the end, what gives us hope and optimism is that music matters, perhaps now more than ever. Music is often the hub of your smart phone experience, it is the backbone and soundtrack to many TV shows, it is the focal point of conversation in social media. Of the top ten most followed people on Twitter, seven are music artists. Of the top six videos on YouTube, five are music. Music remains a centrifugal force in culture and commerce, and it's only going to get stronger. It's worth creating, and it's worth protecting.

My conclusion? The music industry is on an exciting and promising trajectory right now. Whether you are with a major label or an indie, or you just have a dream and are trying to find an audience on your own, online and mobile services give artists and the people who invest in them a chance to succeed. And a broad array of authorized music services are giving music fans what they want, and how they want it.

We've still got a lot of work to do to get piracy under control, and we need the help of other industry sectors in the Internet ecosystem to get there, especially search engines, who have yet to dedicate themselves like others in that ecosystem have to protecting against theft. But we're seeing the results of more than a decade of adapting to the digital environment finally start to come to fruition. Our companies have been working with an enormous number of partners to embrace every opportunity, to license every viable kind of digital music service and make them attractive and successful, and they deserve great credit for the transformation of their business models. We will continue to work with anyone dedicated to the legal consumption of music. We believe that we have a vibrant digital future, and relying on your help to protect our rights, we look forward to marketplace solutions to get there.

Thank you.

In: Future of Audio and Video Markets ... ISBN 978-1-62257-764-4
Editors: C. Powell Long, F. McDonald

Chapter 14

STATEMENT OF JEFF SMULYAN, CHAIRMAN, PRESIDENT AND CEO, EMMIS COMMUNICATIONS CORPORATION. HEARING ON "THE FUTURE OF AUDIO"*

Good morning Chairman Walden, Ranking Member Eshoo, and Subcommittee members, my name is Jeff Smulyan. I am the Chairman, President and CEO of Emmis Communications Corporation, which owns and operates 20 radio stations in six markets in the United States, as well as radio stations in Slovakia and Bulgaria. Emmis also has a publishing division, which publishes the *Texas Monthly* and other city and regional magazines.

Thank you for the opportunity to speak with you today about the valuable services provided by local radio stations like those operated by Emmis. My testimony will focus on the public interest benefits of extending the reach of those services through mobile phone devices. Unlocking the mobile phone market would provide the millions of mobile phone subscribers with greater access to the free news, public affairs and entertainment programming provided by local radio broadcasters. Perhaps most importantly, the inclusion of activated radio chips in mobile phones would expand access to the critical public safety information that Americans have come to expect from their local radio broadcasters during times of emergency.

* This is an edited, reformatted and augmented version of the testimony given on June 6, 2012 before the House Committee on Energy and Commerce.

Local Radio Serves the Public in Many Ways, Particularly during Emergencies

With a growing weekly audience of 241 million listeners age 12 and older,[1] local radio's reach is unparalleled. Radio's strength and popularity benefits recording artists and record labels through free promotion, but more importantly, serves listeners across the country in myriad ways.

Radio is the primary audio source for local news and information, political discourse, music and other entertainment. Despite difficult economic conditions, and an increasing number of alternatives for consumers' attention, radio stations produce and provide many hours of live, local programming, including news and information, every day, along with national news and public interest programming. Indeed, studies have shown that those who listened to news and to discussions about campaigns on the radio showed greater interest in political campaigns and were more likely to hold specific opinions on political issues and to be more aware of candidates' positions on policy issues.[2] And, as the Federal Communications Commission's ("FCC") *Future of Media Report* recognized, the importance of radio to the public discourse has heightened even more in recent years with the rise of all news/talk formats.[3]

Beyond providing a wide array of programming, radio stations are committed to serving their local communities in other tangible ways. The average radio station airs hundreds of Public Service Announcements each year, the majority of which pertain to local community issues, and provide critical support for the fundraising efforts of local charities, other community organizations and disaster recovery projects.

Moreover, radio broadcasting is, and will continue to be, the optimal audio method for reaching mass audiences during emergencies. America's local radio broadcasters provide a powerful combination of nationwide ubiquity and local journalism that has served communities as the primary audio source of information during emergencies and disasters for nearly 100 years. Despite significant advances in communications during that period, local radio remains irreplaceable as a means to inform the public. Through our unique role as the backbone of the Emergency Alert System ("EAS"), as well as the emergency journalism provided by local stations, radio helps protect lives and property during severe weather conditions and other disasters.

We embrace our role as "first informers" during times of emergency. Federal Emergency Management Agency Chief Administrator Craig Fugate recognized this when he instructed Americans to turn to their local radio stations for critical information as Hurricane Irene approached the East Coast last year.[4] Radio stations took similar measures to aid citizens during the rash of tornadoes that have devastated the nation's mid-section the past few years, the floods in North Dakota and storms in Massachusetts in 2011, and numerous other emergency situations.[5] In addition, radio stations helped create and are active partners with law-enforcement agencies in the approximately 120 local, regional and statewide AMBER Plans across the nation. Since the program began in 1997 in the Dallas, Texas area, the AMBER Plan has been credited with successfully returning 584 abducted children.[6]

The "one-to-many" architecture of broadcasting provides powerful signals that blanket communities. It is the most robust, reliable mode of delivery of information to a mass audience, especially during emergencies, when wireline and wireless networks (including Internet access) can be quickly overwhelmed by a surge in traffic. Radio is infinitely scalable to additional users, and because radio tuners rely on over-the-air signals, there is no risk of network congestion. During a disaster, this superior reliability could be the difference between life and death, and after major disasters, radio signals may be the only connection available to citizens in harm's way. Also, during major disasters, the electrical grid often goes down, rending cellular networks useless. However, most broadcasters have back-up emergency generators that allow them to be the only point of connection and information for many Americans.

Access to Broadcast Radio via Mobile Phones Would Benefit Consumers and Enhance Public Safety

For all the reasons discussed above, it is imperative that Americans be able to access radio service over the most ubiquitous, widely-carried portable device, the mobile phone. The consumer benefits of incorporating radio reception into mobile phones are many and varied. Americans will be able to listen to their favorite local radio stations wherever they are, over their mobile phones. This also means mobile acccss to local news, entertainment, weather and traffic information, and potentially life-saving information during emergencies.

Let me be clear: despite the continued claims of wireless industry advocates, broadcasters are *not* seeking a mandate that radio chips be included in mobile devices. Rather, we have worked to educate policymakers on the benefits of expanding the availability of radio-enabled mobile phones. We have also worked for many years to incentivize wireless operators to expand consumer options for radio-enabled devices. For example, last month Emmis Interactive, iBiquity Digital Corporation and Intel (with financial support from NAB Labs) announced the development of a new energy efficient, affordable, state-of-the-art HD radio chipset. This new technology offers song tagging features and delivers an enhanced, interactive radio experience, for the mutual benefit of wireless carriers and radio stations, as well as their subscribers/listeners and advertisers.[7] We are in the process of presenting this exciting new opportunity to major wireless carriers.

Given the popularity and value of radio service, access to broadcast radio could be, and should be, an attractive feature in mobile devices. To date, however, wireless operators have declined to produce radio-enabled mobile phones on a scale commensurate with general consumer demand for radio.

Indeed, recent evidence demonstrates the public's strong interest in access to radio via mobile phones. According to a recent Harris Interactive poll, an increasing number of Americans who own a mobile phone would consider paying a small, one-time fee to access their favorite local radio stations on their phone.[8] Specifically, 81 percent responded positively to this inquiry, compared to 76 percent in 2010. For mobile phone owners with children at home, this figure is 85 percent, up from 79 percent in 2010, and retirees who favor radio chips in mobile devices rose to 76 percent from 66 percent in 2010. The Harris survey documented that 70 percent of mobile phone owners believe that having a radio built into their phones would be "very" or "somewhat" important, and 76 percent of adults and 86 percent of 18-34 year-olds would use a radio built into their mobile phones. Assertions by the wireless industry as to an alleged lack of demand for radio in mobile devices are thus contrary to evidence.

Consumer demand for radio-enabled mobile devices is further demonstrated by looking at the global market. For example, a 2008 study from TNS found that 45 percent of mobile users in Latin America and Asia cite AM/FM radio as one of their top three reasons for purchasing a mobile phone — making it more popular than mobile Internet access, texting and a camera function.[9] It is estimated that over 800 million radio-enabled mobile phones are available on the global market. Nokia alone has sold more than 700 million mobile devices with radio capability. In the United Kingdom, approximately 56 percent of mobile phone models offered by the two primary wireless carriers include radio as a feature. Compare that to the U.S. market, where a recent survey found that only 21 percent of AT&T, 21 percent of T-Mobile, and seven percent of Verizon's mobile phone models offer broadcast radio as an available feature.

In the global market, mobile phone carriers generally do not exercise gatekeeper control over which features are included in mobile devices. There, manufacturers create phones with a variety of options, and consumers choose their devices based on which features they value. Consumers separately choose their carrier based largely on price and network reliability.

In stark contrast, the U.S. mobile phone market is characterized by exclusive contracts between consumer electronics manufacturers and mobile phone carriers. It is incumbent upon manufacturers to conform to the device specifications desired by carriers if they want to meaningfully compete in the market.[10] This is clearly shown by the fact that many of the *same* models of phones include radio capability when sold overseas, but not when sold in this country. In most cases, the radio chip is present in both identical models, but the chip is simply not activated in the United States. As a result, the percentage of mobile phones with radio reception in the United States significantly lags the global market.

Moreover, it is needlessly difficult for consumers to identify the few device models that do include activated radio chips. Radio is not typically featured, or even listed, as an available search option on wireless carriers' phone purchasing websites. Verizon's website, for example, allows consumers to research phones by selecting and searching for any combination of 18 various features; however, Verizon does not include free, over-the-air radio on that long list of features.[11] The story is the same at mobile phone retail stores, where radio is rarely indicated as an available feature on the display cards for mobile devices. In-store salespersons and telephone customer service representatives are largely unaware as to which of their own devices are radio-enabled. Indeed, while they are often surprised to learn that a particular phone is equipped to receive free, over-the-air radio, they are clearly aware of, and able to inform consumers about, streaming, data-based radio apps, like Pandora.com.[12] Some industry experts observe that mobile phone providers would rather reap the revenue of data-intensive, fee-based streaming apps than offer consumers a free and local audio alternative.

The control of U.S. wireless carriers over the manufacture of mobile devices ultimately results in consumers being unable to access radio's free, local news, information and entertainment programming via their mobile phones. More seriously, this lack of access could jeopardize the lives and property of Americans by restricting the availability of critical information during emergencies. Including and activating radio chips in mobile phone devices, and making them widely available to consumers, would substantially increase the accessibility of important emergency information, and keep Americans safer.

Both members of Congress and the FCC have recognized this potential. In February, members of the Congressional Black Caucus called on the FCC to hold a hearing to explore the benefits of including activated radio tuners in mobile phones.[13]

Following a rash of severe weather and tornado outbreaks in the summer of 2011, former FCC Commissioner Michael Copps said, "We share a duty to think creatively about how we can arm consumers with additional ways to communicate during disasters. . . I think the time is here for a thorough, calm and reasoned discussion about FM chips in handsets."[14]

I recognize that the wireless industry is deploying a text-based Commercial Mobile Alerting System ("CMAS") system (also known as "Personal Localized Alerting Plan" and "Wireless Emergency Alerts"). But, the limitations of that system restrict emergency warnings to a 90-character text that lacks the critical details routinely provided by local radio stations during an emergency.

Thus, important, lifesaving information, such as the location, path and expected impact of a severe storm or other disaster, and public safety instructions concerning shelter and evacuations typically is not provided on CMAS. Radio stations also are able to provide valuable post-disaster information (e.g., where to obtain emergency provisions or medical care, how to assist in disaster relief, etc.). Thus, radio chips in mobile handsets, in conjunction with CMAS texts, would provide vital emergency warning redundancy, and offer "one-stop shopping" for both initial alerting and the critical, potentially life-saving emergency information that Americans expect from local radio broadcasters.

The FCC has specifically recognized the benefits of a CMAS-radio chip combination in mobile phones. As noted in the FCC's *Future of Media Report*, "FM chips in mobile devices can provide a number of benefits to consumers."[15]

During emergencies, broadcast chips would greatly assist listeners in obtaining information and, as the *Future of Media Report* noted, could enhance other emergency notification services. "[A]fter getting a short text about the emergency, [consumers] could tune into radio news broadcast for more information (particularly if congestion on mobile networks or power outages make it hard to get on the Internet)."[16]

CONCLUSION

In the end, this is a matter of consumer access. Americans deserve better choices and should be free to choose mobile phones with activated radio chips. Both the demonstrated demand for radio-enabled mobile phone devices, and local radio's role as a lifeline service during times of crisis, are considerations that this Subcommittee should take into account as it explores this issue.

Thank you again for the opportunity to speak with you today.

End Notes

[1] *Radio's Audience Continues to Grow*, Arbitron Newsroom (Dec. 5, 2011), available at http://arbitron.mediaroom.com/index.php?s=43&item=793 .

[2] D. Drew and D. Weaver, "Voter Learning in the 2004 Presidential Election: Did the Media Matter?," 83 *Journalism & Mass Communication Quarterly* 25, 38 (Spring 2006); S. Kim, D. Scheufele and J. Shanahan, "Who Cares About the Issues? Issue Voting and the Role of News Media During the 2000 U.S. Presidential Election," *Journal of Communication* 103, 11-12 (March 2005).

[3] *The Information Needs of Communities*, Report, Federal Communications Commission, at 66 (July 2011), available at http://www.fcc.gov/info-needs-communities ("*Future of Media Report*") (noting that news/talk radio serves an important function in a democracy by giving voice to millions who use the medium to express their opinions).

[4] *See* http://www.cnn.com/video/#/video/bestoftv/2011/08/25/exp.am.craig.fugate.cnn .

[5] *See* Ann Marie Cummings, *Broadcasters: America's "First Informers*," National Association of Broadcasters (Jan. 31, 2012), available at http://nabroadcasters.wordpress.com/2012/01/31/broadcasters-americas-first-informers/ .

[6] *See* http://www.missingkids.com/missingkids/servlet/PageServlet?LanguageCountry=en_US&PageId=4319 (last visited May 31, 2012).

[7] *See* Leslie Stimson, *HD Radio, Cell Ecosystem Hailed*, Radio World (Apr. 17, 2012), available at http://www.nabshowdaily.com/2012/TuesdayEdition/127866 .

[8] *See* http://www.nab.org/documents/newsRoom/pdfs/050812_Radio_Chips_Cellphones_Survey.pdf.

[9] *Mobile Reviving Radio Listening: TNS Survey*, IndianTelevision.com (Mar. 5, 2008), available at http://www.indiantelevision.com/headlines/y2k8/mar/mar50.php , stating that "in markets like India, a mobile phone without FM radio is difficult to sell."

[10] *See Study of the Potential for FM Radio to be a Universal Feature on Cellular Handsets*, Dr. Joseph S. Kraemer, Director, Law and Economics Consulting Group, at 24-25 (May 28, 2008), available at http://www.nabfastroad.org/Reports/FMRadioFeatureCellularHandsets052808.pdf .

[11] *See* http://www.verizonwireless.com/b2c/store/controller?item=phoneFirst&action=viewPhoneOverviewByDevice&deviceCategoryId=1 (last visited June 1, 2012).

[12] *See Study of FM Radio-Enabled Handsets in the U.S.*, The Insight Research Corporation (Sep. 2010), available at http://www.nabfastroad.org/Reports/FM-Enabled_Cell_Phones_9-16a.pdf .

[13] Letter from the Honorable Yvette D. Clarke (D-NY), *et al.*, to FCC Chairman Julius Genachowski (Feb. 3, 2012), available at http://www.radiorocksmyphone.com/ClarkeFMChipLetter-Final2012.pdf

[14] Remarks of Commissioner Michael J. Copps, Workshop/Webinar on Proposed Extension of Outage Reporting and on Network Reliability and Continuity (Sep. 8, 2011), available at http://hraunfoss.fcc.gov/edocs_public/attachmatch/DOC.

[15] *Future of Media Report*, at 309.

[16] *Id.*

In: Future of Audio and Video Markets ... ISBN 978-1-62257-764-4
Editors: C. Powell Long, F. McDonald

Chapter 15

STATEMENT OF STEVEN NEWBERRY, PRESIDENT AND CEO, COMMONWEALTH BROADCASTING CORP. HEARING ON "THE FUTURE OF AUDIO"*

Good morning, Chairman Walden, Ranking Member Eshoo and members of the Subcommittee, and thank you for inviting me to testify today. My name is Steve Newberry, and I am President and CEO of Commonwealth Broadcasting Corporation, which operates 24 radio stations in Kentucky. I am testifying today on behalf of the free, local, over-the-air radio members of the National Association of Broadcasters.

INTRODUCTION

For ninety years, broadcast radio has impacted the lives of Americans in many beneficial and significant ways. Radio broadcasters inform, educate and alert our listeners to important events, topics and emergencies. We introduce them to new music. We entertain them with sports, talk and interviews. We are local, involved in our communities and proud to serve the public interest.

Technological changes over the past decade have led to exciting new developments in the radio industry. Streaming, podcasting, HD radio, mobile devices and other new platforms present both opportunities and challenges for

* This is an edited, reformatted and augmented version of the testimony given on June 27, 2012 before the House Committee on Energy and Commerce.

radio broadcasters. Digital distribution is still only a small part of overall audio consumption, but it is providing innovative ways for us to reach and serve our listeners.

One thing that has not changed is America's love for radio and all that the word "radio" embodies. More people are listening to broadcast radio than ever before. According to 2012 Arbitron data, approximately 241 million persons aged 12 and over listen to radio each week, a number that continues to increase.[1] Indeed, that number is some 2 million higher than in Arbitron's 2010 study, which found at that time, that broadcast radio reached more than 93 percent of persons aged 12+ each week and about 91 percent of young listeners aged 12-17.[2] This last percentage is particularly interesting since teens in this age group are the most accustomed to using new technologies and media platforms. The data also show radio's universal appeal, with more than 93 percent of African Americans and 95 percent of Hispanic persons tuning in to radio each week.[3]

Given this evidence of broadcast radio's continuing appeal, I am not at all surprised that new digital music services endeavor to style themselves as "radio." They want to claim our heritage, but the concept and reality of the radio industry that I represent before you today is much more than the mere audio transmission offered by many services. We are part of the fiber of our local communities, and we intend to stay that way.

The radio industry looks forward to a robust future that embraces the fundamental nature of broadcasting, as well as new opportunities arising from evolving digital technologies. In that regard, however, broadcast radio today faces numerous challenges, and I would like to mention just a few of them here.

BROADCASTERS SHOULD NOT BE SUBJECT TO A NEW PERFORMANCE TAX

Efforts to encourage Congress to establish a new performance fee, what we call a "performance tax," come at a volatile time for both the radio and recording industries. Both industries are fighting intense competition for consumers through the Internet and other new technologies, and both industries are experiencing changes to their traditional business models. These technological and marketplace developments, however, do not justify a government-mandated shift in the relationship between radio and the recording industry.

From the very first days of commercial radio, broadcasters and the music and recording industries have enjoyed a well-balanced relationship that has benefited all parties. Record labels and performing artists profit from the free exposure provided by radio airplay, while local radio stations receive revenues from advertisers that purchase airtime to sell their products and services. Despite the many dramatic changes that have occurred in the digital music industry over the past decade, this interdependent relationship between radio and the music and recording industries remains fundamentally the same.

What has changed is the financial dominance of the four (perhaps soon to be three) major record labels.[4] Digital audio transmission services abound, offering nearly unlimited opportunities for consumers to listen to music on-demand, to make digital copies of songs, and to create personalized listening experiences for themselves and others.

In contrast, while making many technological improvements, radio broadcasting retains the same basic character that it has had for decades. It is local. It is free to listeners. It is supported by commercial advertising. Local stations use on-air personalities and DJs to differentiate their programming, including by commenting on the music they play. While increasing, there are not an unlimited number of radio stations in the U.S., and stations still expose listeners to new and varied songs by choosing what to play. In addition, radio is characterized by its public service to local communities and is subject to numerous Federal Communications Commission ("FCC") restrictions and obligations.

As I mentioned, many digital audio transmission services are eager to associate themselves with radio's rich history, consumer familiarity and affection. They often style themselves as offering "radio" services. But simply marketing digital audio transmission services as "radio" does not make them so. In fact, in 1995 and 1998, Congress recognized the vast differences between digital audio transmission services and local radio when it created a limited digital sound recording performance right for those new services that diverged so dramatically from the nature of traditional radio.

Prior to 1995, U.S. copyright law did not recognize *any* public performance right for owners of sound recordings. The narrow digital performance right Congress created in 1995 was intended to address very specific concerns about copying and piracy issues. This limited right did not attach to a wide variety of recorded music, including radio, hotels, restaurants, bars, nightclubs, sporting arenas, shopping malls, retail stores, health clubs, etc. As explained in the Senate Report accompanying the Digital Performance Rights in Sound Recordings Act of 1995, "The underlying rationale for creation of this limited right is grounded in the way the market for prerecorded music has developed, and the potential impact on that market posed by subscriptions and interactive services – but *not* by broadcasting and related transmission."[5] Consistent with Congress's intent, the DPRA expressly exempted non-subscription, non-interactive transmissions, including "non-subscription broadcast transmission[s]" – transmissions made by FCC-licensed radio broadcasters, from any sound recording performance right liability.[6]

Despite the advent of new technologies and digital audio transmission services that permit sophisticated user manipulation of music in on-demand and customized ways, the impact of the promotional value of traditional local radio remains strong. The fact that consumers have new ways in which to locate and obtain music does not diminish the value of over-the-air radio's marketing and promotion to the recording industry.

In the new, fragmented world of the digital environment, in which millions of bands are vying for the attention of hundreds of millions of fans, on millions of websites, one of radio's greatest strengths is that it cuts through the clutter. Radio exposes listeners to new music and new artists and drives listeners to the websites where their desire for the music that they heard can be monetized.

Congress has repeatedly rejected calls by the recording industry to impose a tax on the public performance of sound recordings. In explaining its refusal to impose new burdens on FCC-licensed terrestrial radio broadcasters, Congress identified numerous features of radio programming that place such programming beyond the concerns that animated the creation of the limited public performance right in sound recordings. Specifically, over-the-air radio programs (1) are available without subscription; (2) do not rely upon interactive delivery; (3) provide a mix of entertainment and non-entertainment programming and other public interest activities to local communities;[7] (4) promote, rather than replace, record sales; and (5) do not constitute "multichannel offerings of various music formats."[8]

Although NAB vigorously opposes the imposition of a Congressionally-mandated performance tax as it has been set forth in previously proposed legislation, in 2010 NAB engaged in discussions with the recording industry in a good faith effort to resolve this issue in the best interests of both radio and the music industry. NAB's Board of Directors subsequently endorsed a thoughtfully-drafted compromise based on those discussions, which was peremptorily dismissed by the recording industry without any constructive comment.

Radio broadcasters continue to oppose any performance tax like those previously proposed because such a tax would undermine local station viability and harm the public's free radio service. As local radio broadcasters have demonstrated on many occasions, stations serve the public interest by airing local and national news and public affairs programming and a variety of other local programming that serves the needs and interests of their audiences, including weather, emergency information, sports, religious and other community-oriented programming.

It goes without saying, however, that maintaining this high level of local programming and other services requires radio stations to be economically sound. Only competitively viable broadcast stations sustained by adequate advertising revenues can serve the public interest effectively and provide a significant local presence. As the FCC concluded two decades ago, the radio "industry's ability to function in the 'public interest, convenience and necessity' is fundamentally premised on its economic viability."[9] Anyone concerned about the service of radio stations to their local communities and listeners must necessarily be concerned about these stations' abilities to maintain their economic vibrancy in light of new fees that could be levied though the creation of a new performance tax. Clearly, these local and community services could be jeopardized by imposition of new fees, especially during difficult economic times.

Broadcasters Urge the FCC to Implement Low Power FM Legislation Consistent with Congress's Balanced Approach

Radio broadcasters are also addressing questions about the relationship between low power FM radio ("LPFM") and full power FM stations as the FCC implements the Local Community Radio Act of 2010 ("LCRA"). My view of the relationship between full-power radio and LPFM is that both services provide value to the American public, and both should be preserved and promoted through appropriate interference protections.

Broadcasters worked closely with Congress to help craft the LCRA in a manner that reasonably balances the needs of both full power FM stations and LPFM stations. The LCRA reflects our common understanding that both full-power FM and LPFM stations serve the public interest, albeit in different, but complementary, ways.

Full power radio broadcasters provide a free, over-the-air service that reaches virtually every household in America, keeping local communities – and your constituents – informed and connected. Local broadcast stations are committed to providing a wealth of local and national news, public affairs programming, political information, vital emergency information, music, sports and other entertainment programming. These broadcasters are also committed to community service as evidenced by billions of dollars annually of free air time devoted to public service announcements, and monies raised for charities and other local causes.

Broadcasters' commitment to public service is particularly evident in times of crisis, such as the ongoing wildfire in New Mexico, and the recent tornado outbreaks in Missouri and Arkansas, during which radio stations delivered life-saving information around the clock, commercial free. Even Federal Emergency Management Agency Chief Administrator Craig Fugate recognized broadcasters' unique role during disasters when he instructed Americans to turn to their local radio station as Hurricane Irene approached the East Coast last year. In addition, through our participation in the Emergency Alert System (EAS) and additional coverage of emergencies, full power radio broadcasters help save lives with extensive, timely emergency information. Coordination with local law enforcement via Amber Alerts has led to the recovery of 584 abducted children.[10] In fact, the Amber Plan was originally created by the Association of Radio Managers with the assistance of law enforcement agencies in the Dallas/Ft. Worth area.

The service that LPFM stations provide is also valuable. LPFM can serve very localized, niche audiences. LPFM stations provide a hyper-local signal, sometimes covering an area no more than a few miles in diameter, with niche programming serving limited sections of a market. Indeed, the FCC created LPFM to "serve very localized communities" and to allow small groups and organizations, such as schools and churches, to provide programming.[11] LPFM stations simply cannot provide the same kinds of community service – especially during emergencies – that full power stations provide. Nor should it be expected to, given their geographic constraints. In addition, as a noncommercial service serving very small geographic areas and discrete audiences, LPFM stations lack the resources to provide the extensive community-wide service offered by full power stations. Few LPFM stations broadcast 24 hours a day, like most full power radio broadcasters.

Nevertheless, both full-power radio and LPFM stations serve the public interest, and the value of each service, as well as their differences, are clearly recognized in the LCRA. The LCRA seeks to balance the dual goals of providing additional licensing opportunities for LPFM stations while preserving the audio quality of full-service FM stations. Specifically, the LCRA eliminated third-adjacent channel minimum distance requirements between LPFM and FM stations, while reaffirming the interference protections of FM stations by prohibiting any changes to the FCC's current second-adjacent spacing requirements. The LCRA permits waivers of these latter requirements, but only under strictly defined, truly unusual circumstances. We have urged the FCC to take a cautious, careful approach to any grant of such waivers, and suggested several reasonable steps designed to ensure that LPFM stations that operate on second-adjacent channels do not inadvertently cause interference to full-service FM stations. For instance, we have proposed that a potentially short-spaced FM station given an opportunity to review the engineering showing of "no interference" required to be submitted by any LPFM applicant seeking a second-adjacent channel waiver. Such a process would help preserve the integrity of the FM band, and also prevent situations in which an LPFM station must cease operations because it has caused interference to a full-service FM station.

Broadcasters have sought to preserve the carefully designed, balanced approach set forth in the LCRA. For example, we have opposed LPFM advocates' proposal to create an entirely new class of LPFM stations that could operate at a maximum of 250 watts. The LCRA is based explicitly on the FCC's current regulations and technical requirements, all of which reflect the long-standing LPFM maximum power limit of 100 watts. Moreover, the LCRA's legislative history, including representations by LPFM advocates themselves, shows that Congress enacted the LCRA based on an understanding of LPFM as a 100 watt service. Accordingly, authorization of 250 watt LPFM stations would not only be contrary to the Congressional basis for the LCRA, but it would change the fundamental, highly localized nature of LPFM service.

Both full power FM and LPFM stations provide valuable service, and the LCRA seeks to preserve and foster both services. Broadcasters believe that if the Commission implements the LCRA in a manner faithful to Congressional intent, thereby preserving the integrity of the FM band, then both services should be able to prosper and serve their listeners going-forward.

The Standard and Procedures Used to Set Streaming Rates Discourage Streaming and Should be Changed

A significant ongoing impediment to broadcasters' ability to innovate in the digital arena is the current rate-setting standard under the statutory licenses for streaming. Rates that have resulted from proceedings by the Copyright Royalty Board ("CRB") under the so-called "willing buyer/ willing seller" standard have been artificially high, to the detriment of both services that wish to stream and the songwriters and performers who would benefit, in the form of increased exposure and royalties, from increased streaming.

Broadcasters favor abandoning the "willing buyer/willing seller" standard and transitioning to the "801(b)(1)" standard for setting streaming royalty rates. The 801(b)(1) standard (so named because it is found in that section of the Copyright Act) has effectively, efficiently and equitably balanced the interests of copyright owners, copyright users and the public for decades, in various contexts and proceedings.[12]

As currently codified, this standard considers the interests of all stakeholders and the public, recognizes the value of all contributions of licensors and licensees and has long been accepted and ratified by Congress. It reflects a Congressional intent not to set rates so onerous that they would stifle new businesses and uses of creative works, or disrupt structure of existing industries. The 801(b)(1) criteria are particularly appropriate where, as now, there are essentially four companies controlling the majority of the distribution of sound recordings.[13]

The "willing buyer/willing seller" standard was perhaps most obviously inadequate when it led to rates for the 2006-2010 license period (set by the CRB) that were so egregious that webcasters directly appealed to Congress. Passage of the Webcaster Settlement Acts of 2008 and 2009 provided an opportunity to negotiate more appropriate arrangements with the recording industry.

Recent developments have further illustrated the dysfunction of the current rate setting procedures. The constitutionality of the appointment of the CRB itself has recently been called into question with an appeal before the D.C. Circuit Court of Appeals. And an additional complication to the broken CRB/SoundExchange system came when SiriusXM recently filed a lawsuit against SoundExchange and A2IM (American Association of Independent Music) claiming antitrust violations. This suit essentially alleges that

SoundExchange and A2IM conspired to prevent SiriusXM from negotiating direct licenses with musicians (which would take music out of the statutory royalty scheme administered by the Copyright Royalty Board).

In short, the royalty rate setting process has become a royal mess, and an opportunity to remedy that process would be embraced by all who stream music. NAB would welcome an opportunity to discuss reform of this dysfunctional process in greater detail.

CONCLUSION

Thank you for the opportunity to testify before the Subcommittee. I believe radio broadcasting will remain strong, and our commitment to our local communities will keep our industry vibrant.

End Notes

[1] *The Infinite Dial 2012, Navigating Digital Platforms*, Study, Arbitron Inc. and Edison Research, available at http://www.arbitron.com/study/digital_radio .

[2] *More Than 239 Million Listen to Radio Every Week According to the Arbitron Radar 105 Report*, News Release, Arbitron Inc. (June 15, 2010), available at http://arbitron.mediaroom.com/index.php?s=43&item=693.

[3] *Id.*

[4] Universal Music Group's proposed merger with EMI Music (subject to U.S. and E.U. regulatory approval) will give Universal roughly 40% of the U.S. market for recorded music.

[5] S. Rep. No. 104-128 (1995) at 17 (emphasis added).

[6] 17 U.S.C. §114 (d)(1)(A).

[7] Radio broadcast stations provide local programming and other public interest programming to their local communities. In addition, there are specific requirements that do not apply to Internet-only webcasters. *See* 47 U.S.C. §§ 307, 309-10 (1998). *See, e.g.,* 47 C.F.R. § 73.352(e)(12) (requiring a quarterly report listing the station's programs providing significant treatment of community issues); 47 U.S.C. . § 312(a)(7) (requiring a station to allow reasonable access to federal candidates); 47 U.S.C. § 315(a) (requiring a station to offer equal opportunity to all candidates for a public office to present views, if station affords an opportunity to one such candidates); 47 C.F.R. § 73.1212 (requiring identification of program sponsors); *id.* § 73.1216 (providing disclosure requirements for contests conducted by a station); *id.* § 73.3526 (requiring maintenance of a file available for public inspection); *id.* § 73.1211 (regulating stations' broadcast lottery information and advertisements).

[8] S. Rep. No. 104-128 (1995) at 15.

[9] *Report and Order*, 7 FCC Rcd 2755, 2760 (1992).

[10] *See* http://www.missingkids.com/missingkids/servlet/PageServlet?LanguageCountry=enUS&PageId=4319 (last visited May 31, 2012).

[11] Report and Order, *Creation of a Low Power Radio Service*, 15 FCC Rcd 2205, 2208, 2213 (2000).

[12] Instead of determining rates for the a statutory license through a hypothetical marketplace, 17 U.S.C.§ 801(b)(1) sets forth four objectives to be considered: "(A) To maximize the availability of creative works to the public; (B) To afford the copyright owner a fair return on his or her creative work and the copyright user a fair income under existing economic conditions; (C) To reflect the relative roles of the copyright owners and the copyright user in the product made available to the public with respect to relative creative contribution, technological contribution, capital investment, cost, risk, and contribution to the opening of new markets for creative expression and media for their communications; (D) To minimize any disruptive effect on the structure of the industries involved and on generally prevailing industry practices."

[13] Significantly, the 801(b)(1) standard has a history of yielding results that are more equitable, effective, and predictable than those supposedly based on market or "fair" value. The relatively uncontroversial and unchallenged results demonstrate the equity of the standard. In each of the four proceedings that have occurred under the 801(b)(1) standard since 1976 (two involving the recording industry as licensor and two involving the recording industry as licensee), not only have the royalty awards been upheld by the courts, but none of the parties have felt compelled to ask Congress to remedy any of the determinations.

In: Future of Audio and Video Markets ... ISBN 978-1-62257-764-4
Editors: C. Powell Long, F. McDonald

Chapter 16

STATEMENT OF TIM WESTERGREN, CHIEF STRATEGY OFFICER AND FOUNDER, PANDORA. HEARING ON "THE FUTURE OF AUDIO"*

Good morning Chairman Walden, Ranking Member Eshoo, and Members of the Subcommittee. Thank you for inviting me to testify this morning on "The Future of Audio." My name is Tim Westergren and I am Chief Strategy Officer and Founder of Pandora.

Twelve years ago, after spending my twenties and early thirties playing in rock bands, and composing film scores, I founded a music discovery service to help independent musicians like me find their audiences. I had a vision that by marrying musicology and technology, I could solve the great problem of discovery - how to help listeners find new music to love, and how to bring all these talented artists the attention and careers they deserve.

It took six long years before the Music Genome Project finally found its home in the form of Pandora. We have gone from a cramped studio apartment in South San Francisco, financed through salary deferral, credit card debt, and the extraordinary effort of a talented group of musicians and engineers, to the country's largest Internet radio service, with over 150 million registered listeners and close to 600 employees in offices all across the country. We stream more hours of music every month on Pandora than YouTube streams

* This is an edited, reformatted and augmented version of the testimony given on June 6, 2012 before the House Committee on Energy and Commerce.

hours of video and refer more music sales to iTunes and Amazon than virtually any other service in the world.

Today, more than ever before, the dream of connecting listeners and artists seems within our grasp. Thanks to Pandora's Music Genome Project, and its rich understanding of musical recordings (every song in our collection is hand-analyzed by a trained musician along hundreds of musical attributes per song), Pandora now plays the music of over 100,000 artists (70% of them independent), represented by a catalogue of over a million songs. And over 95% of these songs play every month. Let me repeat that. Over 950,000 unique songs play every month on Pandora. And the music spans hundreds of genres that are rarely heard in any medium – bluegrass, big band, classic country, baroque, klezmer, a cappella, New Orleans jazz, you name it. For most of these artists, Pandora is the only radio play they've ever enjoyed. It is conceivable that this new promotional vehicle, if it continues to grow, may eventually lead to the emergence of a musician's middle class.

We have created a democratic and inclusive form of radio that listeners can access the same way they listen to terrestrial broadcasting. Eight of the world's largest automakers now include Pandora in new models. We are embedded in over 650 consumer electronics devices that enable Pandora to be enjoyed throughout people's homes. We are the second most downloaded iPhone app. Pandora is even built into refrigerators. Long gone are the days when customers accessed internet radio only through their PCs or laptops. In fact, over 70% of our listening now takes place off the computer.

But beneath this exciting transition is a severe and fundamental problem. While Pandora and other Internet radio services compete directly with broadcast and satellite radio for listeners in every place you find music – the home, the car, the office, on the go -we are subject to an astonishingly disproportionate royalty burden compared to these other formats.

All forms of radio pay *songwriters* a very similar percentage of revenue but there are enormous differences in how *performing artists and labels* are compensated.

Last year, on revenues of $274MM, Pandora paid $137MM in performance fees to performing artists and labels, or 50% of revenue. That same year, Sirius/XM, on revenues of $2.74B, paid $205MM, or 7.5% of revenue; and broadcast radio, on revenues of roughly $15B, paid zero. Now I am fully supportive of fair compensation for artists. I'm a musician, and I strongly believe that radio can and should reward musicians for the use of their work – both songwriters AND performers. But this lack of a level playing field is fundamentally unfair and indefensible.

So why is there such a disparity in royalties? The inequity arises from the fact that Congress has made decisions about radio and copyright law in a piecemeal and isolated manner; as each new form of radio transmission was invented, new legislation was passed but only to address the new form. The effect has been to penalize new media and advantage old media when setting the rules for music royalties.

To give you a sense of how absurd the current situation is, consider this example: Drivers of most current Toyota cars can receive AM/FM, satellite and internet radio, all through the vehicle's sound system. If a song is delivered over AM/FM, the associated performing artist and label receive no compensation. If the same song is delivered over satellite, the performing artist and label do receive compensation. But if the song is delivered using internet transmission over 3G or 4G, then the effective percentage of revenue that must be paid by the company delivering the song goes up six fold. It's the same song, same car, and the same sound system.

It is time for Congress to level the playing field and to approach radio royalties in a technology neutral manner. The current rate-setting law is unfair to performing artists, unfair to record labels, and unfair to Pandora and internet radio as we compete every day with broadcast radio and satellite radio for listener loyalty and advertising and subscription revenue. America's obligations to our performers should be shared proportionally by all who use their music. What amounts to a subsidy of entrenched radio media stifles innovation, discriminates against the internet and adversely affects consumers. It is time to unshackle innovation and allow new technologies and new forms of audio consumption to compete fairly.

Thank you for your time and consideration. I look forward to your questions

In: Future of Audio and Video Markets ... ISBN 978-1-62257-764-4
Editors: C. Powell Long, F. McDonald

Chapter 17

TESTIMONY OF CHRISTOPHER GUTTMAN-MCCABE, VICE PRESIDENT REGULATORY AFFAIRS, CTIA. HEARING ON "THE FUTURE OF AUDIO"*

Chairman Walden, Ranking Member Eshoo, and members of the Subcommittee, on behalf of CTIA – The Wireless Association®, I want to thank you for the opportunity to participate in this morning's hearing on "The Future of Audio."

From the vantage point of the wireless industry, we believe the future of audio is mobile, it is personal, it is what the consumer wants it to be.

I want to share with you some of the reasons we feel this way, as well as a few suggestions for how to ensure that our optimistic projection turns out to be true.

As mobile broadband becomes more and more ubiquitous, the mobile platform – whether accessed by phone, by tablet, or by some other connected device – provides a compelling opportunity for artists and entrepreneurs to deliver, and consumers to access, all sorts of audio products. Whether their interests lie in music, in news, in sports, or even in politics, consumers can and will be able to access content of their choosing, tailored to their preferences of the moment.

This shift from a producer and distributor driven model to a consumer driven model will test existing businesses and force them to adapt. We see this

* This is an edited, reformatted and augmented version of the testimony given on June 6, 2012 before the House Committee on Energy and Commerce.

in the way that the music and news gathering businesses, for instance, are being transformed by the twin forces of digitization and disintermediation. This is a challenge for some in the old order, in much the same way that telephony was a challenge for the telegraph, and wireless is a challenge for wired. At the same time, these forces are unleashing a wave of creativity and innovation that is giving rise to new businesses like Pandora, Spotify, and TuneIn, morphing iPods into iPhones, and enabling new services, like Cricket's MUVE, which has greeted enthusiastically by consumers and investors alike.

While many of us who grew up in the age of the LP lament the loss of album art, liner notes, and lyric sheets, we should recognize that this transformation is one that is good for consumers - and good for America. Consumers gain control, as they can choose the time and method by which their content is delivered or accessed. And America wins because it is at the center of this transformation, as home to many of the innovators bringing these new services to market and as the world's undisputed leader in wireless broadband.

So what needs to be done to enable continued American leadership and ensure that the future of audio is, indeed, bright? The short answer is that we must continue to have access to spectrum, which is the highway across which audio traffic moves. We need a light regulatory touch that will enable service and applications providers to experiment with new product offerings and pricing structures. And we need to resist calls to use the legislative or regulatory process to protect or extend the life of business models that find themselves subject to new competitive challenges.

I want to expand a bit on this last point, as I know that some of our friends in the radio industry will use this hearing as an opportunity to reiterate their call for wireless devices to include an FM chip or some other type of commercial radio capability.

To be clear, CTIA is not opposed to including FM capability in wireless devices. Many of CTIA's manufacturers build handsets that include FM chips and each of the largest carriers offers handsets that contain this functionality. Indeed, close to fifty wireless devices today have a FM chip included. But the decision to offer FM capability – or not – should be driven by consumer preference, not government fiat, and for this reason we absolutely oppose a mandate involving FM chips. We take a face value NAB President and CEO Gordon Smith's statements to this Subcommittee that he is not seeking a mandate in this space and are pleased that he agrees with us on this point.

CTIA opposes a mandate in this space for a variety of reasons, but mostly because such a requirement is at odds with the idea that a competitive marketplace does a superior job of delivering products and services that consumers value. Our members provide phones with and without radio capability, just as they provide handsets with or without keypads, and devices with or without Internet access capabilities. This differentiation allows consumers, who often have widely divergent preferences, to find what they want and value, rather than taking any phone they want, as long as it's black and connected to the wall, as was once the norm.

Proponents of an FM chip mandate will try to wrap themselves in the cloak of public safety to justify their request. I urge you to reject this effort, which has been considered and dismissed by a panel charged with ensuring that the mobile platform can augment the emergency alerting capabilities currently provided by television and radio broadcasters.

I had the privilege of serving on the Commercial Mobile Service Alert Advisory Committee when it met to develop recommendations for implementation of the Warning, Alert and Response Network, or WARN, Act. In evaluating the technological options available for executing on the WARN Act's call for a mobile alert system, the CMSAAC considered and rejected the use of technologies like an FM receiver, a NOAA Weather Radio, a paging chip, and a satellite chip, as possible solutions to enabling emergency alerting in commercial mobile devices. As the CMSAAC noted, there are serious technical challenges to using these non-native services in commercial mobile handsets, including providing an antenna that is not integrated with a wired headset and addressing power consumption. Additionally, none of these potential solutions offer the geo-targeting opportunities that are possible with the cell broadcast model that was ultimately adopted and is now available to many consumers.

If some in the radio business want to encourage consumers to buy FM-enabled devices, they are free to do so, but they should not seek to have the government impose their business plan on others. The CMSAAC's conclusion was correct in finding that FM is not appropriate for wireless emergency alerts, and thus decisions regarding the inclusion of FM capability in wireless devices must be driven by consumer preference and market forces. That will produce better outcomes for platform providers, content creators, and consumers, as Congressman Issa and Congresswoman Eshoo have acknowledged in their "Creativity and Innovation Resolution," H.Con.Res. 42, which CTIA is pleased to support.

Mr. Chairman and members of the Subcommittee, thank you for the opportunity to appear on the panel. With the right combination of good spectrum policy and regulatory restraint, reliance on the innovative capabilities of American entrepreneurs, and trust that consumers know what they want, the future of audio is, and should remain, bright.

In: Future of Audio and Video Markets ... ISBN 978-1-62257-764-4
Editors: C. Powell Long, F. McDonald

Chapter 17

TESTIMONY OF CHRISTOPHER GUTTMAN-MCCABE, VICE PRESIDENT REGULATORY AFFAIRS, CTIA. HEARING ON "THE FUTURE OF AUDIO"*

Chairman Walden, Ranking Member Eshoo, and members of the Subcommittee, on behalf of CTIA – The Wireless Association®, I want to thank you for the opportunity to participate in this morning's hearing on "The Future of Audio."

From the vantage point of the wireless industry, we believe the future of audio is mobile, it is personal, it is what the consumer wants it to be.

I want to share with you some of the reasons we feel this way, as well as a few suggestions for how to ensure that our optimistic projection turns out to be true.

As mobile broadband becomes more and more ubiquitous, the mobile platform – whether accessed by phone, by tablet, or by some other connected device – provides a compelling opportunity for artists and entrepreneurs to deliver, and consumers to access, all sorts of audio products. Whether their interests lie in music, in news, in sports, or even in politics, consumers can and will be able to access content of their choosing, tailored to their preferences of the moment.

This shift from a producer and distributor driven model to a consumer driven model will test existing businesses and force them to adapt. We see this

---

* This is an edited, reformatted and augmented version of the testimony given on June 6, 2012 before the House Committee on Energy and Commerce.

in the way that the music and news gathering businesses, for instance, are being transformed by the twin forces of digitization and disintermediation. This is a challenge for some in the old order, in much the same way that telephony was a challenge for the telegraph, and wireless is a challenge for wired. At the same time, these forces are unleashing a wave of creativity and innovation that is giving rise to new businesses like Pandora, Spotify, and TuneIn, morphing iPods into iPhones, and enabling new services, like Cricket's MUVE, which has greeted enthusiastically by consumers and investors alike.

While many of us who grew up in the age of the LP lament the loss of album art, liner notes, and lyric sheets, we should recognize that this transformation is one that is good for consumers - and good for America. Consumers gain control, as they can choose the time and method by which their content is delivered or accessed. And America wins because it is at the center of this transformation, as home to many of the innovators bringing these new services to market and as the world's undisputed leader in wireless broadband.

So what needs to be done to enable continued American leadership and ensure that the future of audio is, indeed, bright? The short answer is that we must continue to have access to spectrum, which is the highway across which audio traffic moves. We need a light regulatory touch that will enable service and applications providers to experiment with new product offerings and pricing structures. And we need to resist calls to use the legislative or regulatory process to protect or extend the life of business models that find themselves subject to new competitive challenges.

I want to expand a bit on this last point, as I know that some of our friends in the radio industry will use this hearing as an opportunity to reiterate their call for wireless devices to include an FM chip or some other type of commercial radio capability.

To be clear, CTIA is not opposed to including FM capability in wireless devices. Many of CTIA's manufacturers build handsets that include FM chips and each of the largest carriers offers handsets that contain this functionality. Indeed, close to fifty wireless devices today have a FM chip included. But the decision to offer FM capability – or not – should be driven by consumer preference, not government fiat, and for this reason we absolutely oppose a mandate involving FM chips. We take a face value NAB President and CEO Gordon Smith's statements to this Subcommittee that he is not seeking a mandate in this space and are pleased that he agrees with us on this point.

CTIA opposes a mandate in this space for a variety of reasons, but mostly because such a requirement is at odds with the idea that a competitive marketplace does a superior job of delivering products and services that consumers value. Our members provide phones with and without radio capability, just as they provide handsets with or without keypads, and devices with or without Internet access capabilities. This differentiation allows consumers, who often have widely divergent preferences, to find what they want and value, rather than taking any phone they want, as long as it's black and connected to the wall, as was once the norm.

Proponents of an FM chip mandate will try to wrap themselves in the cloak of public safety to justify their request. I urge you to reject this effort, which has been considered and dismissed by a panel charged with ensuring that the mobile platform can augment the emergency alerting capabilities currently provided by television and radio broadcasters.

I had the privilege of serving on the Commercial Mobile Service Alert Advisory Committee when it met to develop recommendations for implementation of the Warning, Alert and Response Network, or WARN, Act. In evaluating the technological options available for executing on the WARN Act's call for a mobile alert system, the CMSAAC considered and rejected the use of technologies like an FM receiver, a NOAA Weather Radio, a paging chip, and a satellite chip, as possible solutions to enabling emergency alerting in commercial mobile devices. As the CMSAAC noted, there are serious technical challenges to using these non-native services in commercial mobile handsets, including providing an antenna that is not integrated with a wired headset and addressing power consumption. Additionally, none of these potential solutions offer the geo-targeting opportunities that are possible with the cell broadcast model that was ultimately adopted and is now available to many consumers.

If some in the radio business want to encourage consumers to buy FM-enabled devices, they are free to do so, but they should not seek to have the government impose their business plan on others. The CMSAAC's conclusion was correct in finding that FM is not appropriate for wireless emergency alerts, and thus decisions regarding the inclusion of FM capability in wireless devices must be driven by consumer preference and market forces. That will produce better outcomes for platform providers, content creators, and consumers, as Congressman Issa and Congresswoman Eshoo have acknowledged in their "Creativity and Innovation Resolution," H.Con.Res. 42, which CTIA is pleased to support.

Mr. Chairman and members of the Subcommittee, thank you for the opportunity to appear on the panel. With the right combination of good spectrum policy and regulatory restraint, reliance on the innovative capabilities of American entrepreneurs, and trust that consumers know what they want, the future of audio is, and should remain, bright.

In: Future of Audio and Video Markets ... ISBN 978-1-62257-764-4
Editors: C. Powell Long, F. McDonald

Chapter 18

TESTIMONY OF GARY SHAPIRO, PRESIDENT AND CEO, CONSUMER ELECTRONICS ASSOCIATION. HEARING ON "THE FUTURE OF AUDIO"*

CEA is the preeminent trade association representing over 2000 U.S. consumer electronics companies. Every one of our members directly or indirectly has products or services relying on or conveying sound. At our industry's annual trade show, the International CES, over 156,000 attendees admired more than 20,000 new products, including numerous innovations in audio.

We welcome this hearing as it is gratifying for Congress to focus on something we cherish but rarely discuss in Washington: the spoken word, music and sound. We appreciate this unprecedented opportunity to share some little known facts and views.

First, I'm proud to share that many of our members make loudspeakers and other audio products here in the United States and export worldwide. Audiophiles around the globe recognize and appreciate the phenomenal quality of these U.S. products including Altec Lansing, Atlantic Technology, Audioengine, Bose, Conrad-Johnson, Definitive Tech, Eminence, Harman International, HUMAN Speakers, Infinitiy, JBL, Klipsch, Koss Headphones, Legacy Audio, Leon Speakers, Inc., Martin Logan, McIntosh Laboratory Inc., Misco Loudspeakers, Mitek, Orb Audio, Polk Audio, Rayco Sound Industries,

* This is an edited, reformatted and augmented version of the testimony given on June 6, 2012 before the House Committee on Energy and Commerce.

Inc., Rockford Fosgate, Specialty Technologies, Thiel Audio Products, Triad, Voxx International, Westone Laboratories, Wilson Audio Specialties and Z Box.

Other members, such as Microsoft, Apple and Livio Radio give listeners extraordinary and previously unimagined ways to enjoy music when and where they wish.

Still others, like Pandora, Google and Grooveshark open up a vast new universe of choices and content to music fans.

Indeed, Internet radio is booming. Recent research by TargetSpot found 42 percent of U.S. households with broadband Internet listen to Internet radio. Similarly, research from the NPD Group found 43 percent of the Internet population listened to digital radio broadcasts in 2011 – up from 29 percent in 2009. The Consumer Electronics Association recently found 39 percent of the Internet population listened to online streaming audio content in the last 12 months and 42 percent listed to MP3 files.

Second, our creativity and innovation is reflected by tens of thousands of independent musicians and by the big company members of the RIAA. The more I travel around the world the more I appreciate how we export our culture through the Internet, our movies and our music. The result is that so many in the world want to be like us and also learn or want to learn English. This is an American influence force multiplier and should not be taken for granted.

Indeed, the popularity of American culture and products abroad ties in directly in with innovation, including innovation in audio, both music and word, as our national brand. CEA believes our long term national strategy must be based on our strength in innovation. If we can continue to innovate, our economy will continue to grow and mitigate our tax and spending dilemma. Of course the challenge of innovation, including innovation in technology and music and news distribution, is that innovation always threatens incumbents and they too often come to Washington to protect their legacy business models.

Third, the world of reproduced audio is parallel to the world of energy in that we had only a few choices or sources in the 1960s: radio, the phonograph, the telephone and television. And as with energy, we have rapidly shifted to a growing set of choices with a phenomenal array of devices, products, sources and services. It's not just music and news over tablets and telephones, its HD Radio providing choices and music over standard radio spectrum, its Sirius-XM satellite radio appreciated by over 22 million subscribers, its Pandora, and Spotify. It's Gracenote cataloguing the world's music and offering music based on emotion or by lyrics. It's Livio Radio offering direct Internet access. It's independent musicians raising literally millions of dollars on internet platforms like KickStarter. It's the plethora of Internet choices including YouTube and sharing over Facebook. It's the websites of artists and fan clubs. It's audio books expanding rapidly in scope and service to children, the elderly and the disabled. It's storage devices beginning with the Walkman which morphed several times into the groundbreaking iPod. It's webcasts and podcasts and playlists.

Consumers who wish to get their audio news and entertainment in a variety of ways have an incredible amount of choices and the marketplace has responded with all its glory expanding quickly the diversity of business models, offerings and content created. Paralleling with video, we have entered a golden age of content creation and while some of the big distribution companies may see their markets shrink, entrepreneurs and new creators rapidly jump in and create new services, products and groundbreaking content.

And just like with energy, we suggest an "all of the above" policy, which means no one source should be given preferential treatment over all others. For this reason alone, we do not agree that Congress should take any action favoring broadcast radio over any other source of audio. Currently, Internet, satellite and cable broadcasters pay significant royalties to performers when their songs are broadcast. By contrast, over their air radio broadcasters are not required to pay performers. This situation is a creature of historical accident, and it is simply unjustifiable. Congress should ensure that over-the -air broadcasters do the right thing, and fairly compensate musicians.

Fourth, the phenomenal growth in digital video, the resulting flattening of TV screens and the high quality of the displays have created huge opportunities in and demand for quality audio. Americans buy over 30 million flat panel displays every year. To ensure the optimal viewing experience, consumers are increasingly looking for quality audio products to ensure that the sound matches the stunning HDTV pictures.

Fifth, recent advancements in the science of sound reproduction are changing the world of audio. In the early 1980s when we leapt from analog to digital and the compact disc and digital radio, some complained that music delivered with this new digital technology lacked "warmth". We rarely hear those complaints today. Instead, we only see enhancements. Listeners can hear, in their own personal environment, audio that is identical to what was captured in the mixing room, the sound stage or concert hall. Surround sound provides a more immersive experience that makes one a participant more than a casual observer and communicates the emotion of the moment.

Dolby True HD is now part of a download music service. More the traditional packaged media business is being complemented by electronic delivery. Vudu, CinemaNow, HBOGo, Netflix, Apple TV and Amazon are streaming television and motion picture content to the home accompanied by multi-channel (up to 7.1) surround sound.

Meanwhile, Americans are using a variety of devices to access their digital music. According to CEA market research, over the last year 70% of Americans listened to music on a desktop or laptop computer, and 60% on an MP3 player.

Sixth, although music is the only area where millions still spend hours enjoying creativity from hundreds of years ago, the fact is that our best times are still ahead of us. Social networking allows collaborative music creation and sharing. Imagine how our kids' creativity will be unleashed as they learn music and collaborate with bands on line. Given the high correlation with music education and math development this is a good thing culturally and scientifically for innovation.

More, at the other side of the spectrum, with an aging population, audio solutions that provide amplification and sound clarity will emerge through devices such as phones and headsets. In fact, CEA has initiated a standards project to enable people with hearing impairments to adjust audio presentation for their specific needs. We will rely in audio delivered electronically for health maintenance, socialization and service to remote, aging and other populations at need.

The tremendous recent growth in interest in high quality headphones likely presages further business opportunities for higher quality content, media and delivery methods. This headphone-driven interest in high-quality audio may help give legitimate content providers a critical marketplace edge over low-quality, unauthorized internet downloads. More, the growth in surround sound also is being followed by surround streaming. Companies like Olive, Sonos, Sooloos, Sonore and others are manufacturing servers that include surround sound music playback.

Finally, if there is one area of concern it is what we can do collectively, to educate Americans about the importance of protecting their hearing. Many in my generation attended – and may still attend – concerts where their passion for the music put them too close to loud speakers. We also know that playing music too loud can affect hearing and for this reason since 1981 most reputable headphone makers have added warnings with their headphones about playing music too loud. And CEA has joined with the RIAA, the American Speech-Language-Hearing Association (ASHA) and others in a national campaign aimed at ensuring children and parents learn early to value hearing as a precious sense that must be protected. We don't seek Congressional action other than the "bully pulpit" that concerned public officials can utilize on this important issue.

Clearly, the audio marketplace for music and information is incredibly vibrant. Remarkable innovations are entering the market at a rapid pace, and thriving new business models are being created. The winners are US consumers, who have unparelled access to news and information from a multiplicity of sources.

With all of these advances in audio distribution and consumption, I am sad to report that we are still fighting efforts to mandate the incorporation of old technologies in new products – namely, radio broadcasters' demands that Congress forcibly intervene in the marketplace and require analog radio receivers be installed in a host of digital devices.

The fact is that at least two dozen phones equipped with FM tuners are already on the marketplace for consumers who desire that feature, although surveys indicate that it is not widely used.

Faced with the fact that analog radio in digital phones is not popular with consumers, broadcasters are trying a new argument. They now claim that FM tuners are somehow necessary so that Americans can be alerted in the event of a tornado. This argument defies logic for a number of reasons.

The vast majority of radio stations operate in unattended mode, meaning without people present to manually control the programming. Some stations do have people present during specific periods of time, such as morning drive time, but operate unattended during other periods. When it comes to informing their audiences about time sensitive information unattended stations typically either don't do it at all, or are very slow to get the information out. This is because to operate in unattended mode the programming has to be planned out and recorded well in advance.

This unattended operation is a natural result of technological progress. Many other industries have also embraced unattended operation, such as banks, which have for years allowed us to get money from ATM machines when there is nobody present at the bank. Also, gas stations allow us to fill up our tanks without attendants present.

Unattended operation is great for consumers because it provides all of us with easy access to services whenever we need them. However, if we need urgent, timely service in a crisis an unattended business is usually not much help. If your car breaks down in the middle of the night it will be easy to find working gas pumps at an unattended gas station, but very hard to find a mechanic. And if there is a local emergency in your community it will be easy to find pre-recorded programming on unattended radio stations, but hard to find up-to-date information about the crisis.

Also, in the event of an emergency, people increasing use their mobile phones to access the Internet or platforms such as Facebook and Twitter. The services provide specific and localized information that is often more helpful than a generalized radio update.

Finally, and most important, wireless carriers and the federal government are already rolling out a system to provide geographically targeted emergency alerts.

The Wireless Emergency Alert (WEA) system, deployed this past April, will transmit emergency alerts for severe weather, as well as AMBER alerts for missing children and Presidential Alerts for national emergencies. The text based messages are vastly more effective than FM transmissions because they will instantly be displayed on the phone. A warning sent by analog radio is useless unless you happen to be listening to the radio at the exact moment the radio is transmitted.

CEA opposes not only a mandate for FM chips in cell phones, but also opposes the broadcasters' current effort to require a government study of this issue. The marketplace has shown that Americans are perfectly capable of deciding for themselves what functions and features they want in their smart phones. Wasting taxpayer funds for something as absurd an unnecessary mandate on innovation is the kind of special interest-driven expenditure that frustrates average Americans.

Clearly, broadcasters have lost their historic monopoly on music transmission, and now exist in a more competitive environment. Indeed, research TargetSpot research found nearly half of those surveyed (47 percent) spend less time listening to broadcast radio than they did a year ago. Not surprisingly, the steepest decline in broadcast radio listenership was among "digital natives" - young adults age 18 to 24.

The correct answer for broadcasters, however, is not to beg Congress to protect their historic business model. Instead, broadcasters must do what other industries do when faced with new market entrants – learn to compete smarter and harder.

Indeed, if the broadcasters wish to compete effectively, redoubling their commitment to HD Radio would be a good first step. Delivered over the AM/FM analog spectrum, HD Radio allows over-the-air broadcasters to offer the same digital sound quality and wide array of programming now offered by their digital competitors.

We applaud Ranking Member Eshoo and Congressman Issa for introducing House Continuing Resolution 42 , the "Creativity and Innovation" declaring, among others, that new Government mandate forcing mobile device manufacturers and wireless carriers to include terrestrial broadcast radio tuners in new mobile devices will stifle innovation, competition, and consumer choice. We hope other members of this subcommittee will join in support of the resolution.

Rather than imposing mandates that would restrict choice and impose costs on consumers, Congress should act to make sure that the vital music marketplace continues to thrive.

Today, a growing amount of music is enjoyed on spectrum-dependent mobile technologies. We applaud Congress' efforts to promote the voluntary auctions of underused television broadcast spectrum, and we call on you to ensure that these auctions take place as expeditiously as possible. The result will be more investments, more startups, and more high-skill, high-wage jobs.

Similarly, as an intellectual property (IP) industry, we call on Congress to promote smart IP policies that protect creators while maintaining the dynamism of the Internet economy. While we had disagreed with some in the content industry over the approach taken in the Stop Online Piracy Act (SOPA), we continue to believe that we can reach a consensus on strategic and targeted ways to cut off funding to overseas "rogue" web sites. We also encourage content owners to aggressively offer licenses to legitimate music services that will compete effectively with unauthorized purveyors

Finally, to compete effectively, American audio companies need access to the best, brightest and most innovative workers in the world. We urge Congress to enact strategic immigration reform, removing barriers to highly-skilled workers and entrepreneurs, and ensure that immigrants receiving advanced degrees in US universities can stay in America to create jobs.

Innovation is driving our economy, and nowhere is this truer than the audio industry. Hardware manufacturers, online innovators, and artists are all taking advantage of extraordinary new business opportunities. As always during times of disruptive innovation, incumbent industries come to Congress and request special protections. We urge Congress to ignore these self-interested pleas, and continue to promote a vibrant and dynamic free-market economy that creates investment and jobs. We look forward to working with you in those efforts.

Thank you.

INDEX

A

abuse, 63
accessibility, 108
accounting, 63
adaptation, 76
Administrative Procedure Act, 84
adults, 47, 106
advancement(s), 62, 135
advertisements, 119
African Americans, 112
age, 33, 40, 48, 85, 91, 104, 112, 126, 130, 135, 138
agencies, 100, 105, 116
aging population, 136
American culture, 134
antitrust, 58, 118
arbitrage, 79
arbitration, 3
ARS, 43
Asia, 106
assets, 3, 10, 43
AT&T, 2, 67, 68, 73, 78, 107
attachment, 34
authentication, 39
authority, 40, 42, 43, 49, 52

B

baggage, 77
ban, 83, 85
bandwidth, 35, 37, 52, 89
banks, 137
bargaining, 57
baroque, 122
barriers, 74, 82, 90, 139
barriers to entry, 82, 90
base, 91
benefits, 3, 6, 14, 27, 33, 35, 50, 54, 61, 75, 86, 103, 104, 105, 106, 108
blogs, 44, 63
bounds, 42
Broadband, 50, 74
buffalo, 43
Bulgaria, 103
Bureau of Labor Statistics, 99
business model, 11, 33, 35, 36, 39, 41, 95, 98, 100, 102, 112, 126, 130, 134, 135, 137, 138
businesses, 11, 70, 72, 89, 118, 125, 129
buyer, 118
buyers, 38

C

cable company, 2
cable operator, 2, 3, 4, 6, 11, 13, 17, 34, 36, 54, 55, 59, 63, 64, 72, 74, 75, 77, 79
cable service, 27

cable system, 6, 35, 38, 39, 43, 44, 54, 55, 56, 58, 78, 79, 84
cable television, 6, 22, 73, 74
Cable Television Consumer Protection and Competition Act of 1992, 42, 54, 64
campaigns, 104
candidates, 104, 119
CBS, 5, 63, 67
cell phones, 85, 101, 138
challenges, 53, 57, 62, 75, 95, 111, 112, 126, 127, 130, 131
charities, 104, 116
chicken, 22
children, 14, 105, 106, 116, 135, 137, 138
citizens, 105
City, 70
clarity, 136
clusters, 56
CNN, 34, 71
Colombia, 44
color, 88
commerce, 101
commercial, 5, 40, 52, 67, 84, 113, 116, 126, 127, 130, 131
commercials, 5, 10, 41, 94
commodity, 77
common sense, 99
communication, 62
Communications Act, 2, 4, 5, 6, 36, 43, 60
community(ies), 39, 41, 48, 49, 59, 61, 69, 70, 73, 94, 96, 100, 104, 105, 109, 111, 112, 113, 114, 115, 116, 119, 138
community service, 115, 116
compensation, 4, 54, 55, 56, 57, 61, 88, 89, 90, 94, 95, 122, 123
competition, 2, 11, 14, 20, 27, 31, 33, 34, 35, 36, 39, 40, 53, 54, 55, 56, 58, 59, 60, 61, 62, 72, 77, 78, 83, 112, 139
competitive advantage, 34, 58
competitors, 13, 19, 22, 34, 35, 37, 49, 53, 55, 60, 78, 86, 138
complement, 26, 48
composition, 94
compulsory copyright provisions, 4
computer, 15, 23, 29, 66, 98, 122, 136
confidentiality, 50
connectivity, 22
consensus, 139
consent, 2, 4, 5, 6, 11, 36, 39, 44, 53, 54, 55, 56, 57, 58, 59, 61, 62, 79, 94
constituents, 54, 116
consumer choice, 14, 23, 26, 27, 139
consumption, 75, 83, 102, 112, 123, 127, 131, 137
contour, 3
cooking, 72
copyright, 4, 5, 36, 39, 41, 44, 86, 89, 94, 99, 113, 118, 120, 123
copyright law, 5, 86, 99, 113, 123
correlation, 136
cost, 20, 22, 23, 31, 61, 63, 64, 75, 120
cost saving, 61
Court of Appeals, 21, 83, 118
covering, 58, 116
creativity, 126, 130, 134, 136
criminal activity, 99
cross-ownership, 83, 84, 85
culture, 95, 99, 101, 134
customer service, 107
customers, 6, 9, 10, 17, 38, 55, 57, 67, 74, 75, 76, 122

D

dance, 72
democracy, 109
democratization, 90
Department of Justice, 44, 93, 94
digital technologies, 112
digital television, 61
Direct Broadcast Satellite, 34
direct broadcast satellite service, 6, 17
disaster, 104, 105, 108
disaster relief, 108

disclosure, 119
discrimination, 20, 37
discs, 69
disposition, 55
distortions, 56, 79
District of Columbia, 44
diversity, 3, 14, 36, 51, 53, 60, 62, 72, 85, 135
DOC, 110
DOJ, 43
dominance, 42, 113
dream, 101, 122
dynamism, 139

E

economic downturn, 48
economies of scale, 64
ecosystem, 48, 50, 100, 102
education, 136
electricity, 57
electromagnetic, 6
emergency, 7, 54, 85, 103, 104, 105, 108, 115, 116, 127, 131, 138
emotion, 135
employees, 29, 121
encryption, 23
energy, 106, 134, 135
enforcement, 64, 99, 105, 116
engineering, 117
entrepreneurs, 98, 125, 128, 129, 132, 135, 139
environment, 27, 102, 114, 135, 138
equal opportunity, 119
equipment, 83
equity, 34, 35, 120
Europe, 15
evidence, 3, 19, 99, 106, 112
evolution, 34, 36, 41
exercise, 107
exposure, 14, 113, 118

F

Facebook, 66, 67, 75, 76, 135, 138
fair use, 5
fairness, 22
faith, 7, 30, 40, 57, 63, 114
families, 14, 15, 23, 48
fear(s), 14, 41, 83
Federal Communications Commission, 83, 104, 109, 113
Federal Emergency Management Agency, 104, 116
federal government, 34, 52, 63, 138
fiber, 16, 112
films, 68
financial, 3, 5, 10, 70, 106, 113
financial support, 106
fish, 72
flexibility, 34, 38, 76, 96
floods, 105
food, 10, 91
force, 38, 52, 101, 125, 129, 134
Ford, 64
foreign language, 60
free trade, 44
freedom, 25, 26, 59
funding, 139
fundraising, 104
funds, 138

G

Georgia, 69
government policy, 11
growth, 31, 69, 75, 76, 83, 96, 135, 136

H

handheld devices, 50
health, 113, 136
hearing impairment, 136
hedging, 63
height, 56
high school, 91
history, 41, 42, 66, 113, 117, 120
homes, 2, 14, 15, 23, 47, 72, 88, 122
host, 137
hotels, 113

House, 1, 9, 13, 25, 26, 29, 30, 33, 47, 65, 71, 81, 87, 93, 97, 103, 111, 121, 125, 129, 133, 139
hub, 101

I

ideal, 51
identification, 119
immigrants, 139
immigration, 139
improvements, 113
income, 48, 94, 96, 120
increased access, 61
increased competition, 4, 14, 38, 84
incumbents, 11, 34, 35, 36, 37, 38, 42, 134
India, 109
individuals, 99
industries, 50, 65, 70, 99, 112, 113, 118, 120, 137, 138, 139
inequity, 100, 123
inflation, 57
infrastructure, 37, 98
injury, 90
insertion, 101
integration, 78
integrity, 23, 117
intellectual property, 90, 94, 95, 96, 139
interface, 76
interference, 3, 59, 85, 115, 117
intermediaries, 100
intervention, 2, 3
investment(s), 3, 10, 57, 63, 66, 75, 76, 77, 120, 139
investors, 126, 130
Israel, 43
issues, 2, 27, 48, 83, 88, 89, 101, 104, 113, 119

J

joint ventures, 38
journalism, 57, 104
justification, 18

L

landscape, 26, 27, 33, 34, 60, 76
laptop, 75, 136
Latin America, 106
law enforcement, 116
laws, 14, 35, 40, 41, 49, 53, 96, 99
laws and regulations, 14, 49
lawyers, 99
layoffs, 99
lead, 43, 91, 122
leadership, 126, 130
LED, 33
legislation, 31, 49, 85, 114, 123
light, 4, 43, 57, 79, 83, 84, 115, 126, 130
linear programming, 15
litigation, 93
local community, 104
local television stations, 58, 64
Louisiana, 69
love, 26, 72, 112, 121
lower prices, 34
loyalty, 123

M

magazines, 103
Major League Baseball, 30
majority, 3, 21, 22, 56, 57, 78, 79, 85, 94, 104, 118, 137
management, 29, 48
manipulation, 114
manufacturing, 136
market economy, 139
market share, 2
marketing, 19, 83, 113, 114
mass, 71, 104, 105
matter, 9, 21, 58, 74, 82, 109
MB, 42, 43, 44, 63, 64
media, 25, 34, 35, 36, 53, 60, 68, 84, 101, 112, 120, 123, 136
medical, 108
medical care, 108
memory, 68
merchandise, 96

mergers, 38
messages, 138
methodology, 18
Mexico, 50, 116
Microsoft, 30, 66, 134
middle class, 122
minorities, 48
Missouri, 116
mixing, 135
mobile device, 51, 68, 72, 75, 101, 106, 107, 108, 111, 127, 131, 139
mobile phone, 25, 50, 67, 68, 98, 103, 105, 106, 107, 108, 109, 138
models, 29, 36, 39, 75, 86, 95, 96, 106, 107, 122
modifications, 27
MOG, 98
mole, 99
monopoly, 11, 34, 138
motivation, 27
multichannel video programming service, 77
multimedia, 76
multiplier, 134
music industry, 91, 93, 94, 97, 101, 113, 114
music videos, 94
musicians, 91, 99, 100, 119, 121, 122, 134, 135

N

national identity, 99
national strategy, 134
negative effects, 40
negotiating, 4, 11, 57, 100, 119
negotiation, 3, 40, 96
network congestion, 105
networking, 136
neutral, 123
new media, 123
next generation, 91
NOAA, 127, 131
non-broadcast programmers, 3
North America, 73, 79

O

omission, 84
openness, 34, 37
operations, 117
opportunities, 19, 23, 31, 61, 74, 95, 111, 112, 113, 117, 127, 131, 135, 136, 139
optimism, 101
outreach, 100
oversight, 48, 89
ownership, 17, 36, 51, 83, 84, 85

P

parallel, 134
parents, 82, 137
participants, 83
pay-TV providers, 2, 4, 11
performers, 86, 87, 89, 90, 118, 122, 123, 135
performing artists, 113, 122, 123
permission, 44, 54, 90
permit, 22, 114
phonograph, 134
piano, 94
piracy, 41, 99, 100, 102, 113
platform, 3, 30, 60, 61, 66, 70, 72, 74, 75, 100, 125, 127, 129, 131
playing, 14, 23, 121, 122, 123, 136
pleasure, 14
police, 95
policy, 35, 42, 51, 52, 54, 59, 63, 72, 79, 84, 104, 128, 132, 135
policy issues, 79, 104
policymakers, 39, 40, 42, 57, 106
politics, 72, 125, 129
pools, 69
population, 134
portability, 76
precedent(s), 19, 42
preferential treatment, 135
preservation, 61, 100

President, v, vi, vii, viii, 1, 2, 29, 33, 42, 47, 71, 82, 93, 103, 111, 126, 130, 133
producers, 30, 35, 74
professionals, 87
profit, 33, 113
project, 136
proliferation, 74
property rights, 93, 100
proposition, 3
protection, 39, 99
public affairs, 72, 103, 115, 116
public interest, 14, 21, 22, 23, 33, 49, 100, 103, 104, 111, 114, 115, 116, 117, 119
public officials, 137
public policy, 55, 60, 61
public safety, 57, 103, 108, 127, 131
public service, 22, 54, 57, 61, 113, 116
publishing, 98, 99, 103
pumps, 137

Q

quality production, 26

R

rash, 105, 108
reality, 58, 74, 112
reception, 16, 47, 54, 71, 105, 107
recession, 74
recognition, 54
recommendations, 42, 127, 131
recovery, 104, 116
redundancy, 108
reform(s), 30, 39, 40, 44, 63, 98, 119, 139
regulations, 6, 11, 41, 42, 63, 117
regulatory framework, 77
relaxation, 85
reliability, 75, 105, 107
relief, 6, 19
rent, 73, 83
reproduction, 94, 135
requirements, 7, 13, 40, 48, 59, 60, 64, 85, 117, 119
resistance, 17
resolution, 89, 101, 139
resources, 66, 75, 95, 116
response, 13
restaurants, 113
restrictions, 4, 7, 36, 74, 79, 83, 113
retail, 68, 107, 113
revenue, 4, 19, 20, 39, 56, 69, 94, 100, 107, 122, 123
rights, 4, 5, 15, 21, 23, 34, 40, 41, 54, 55, 58, 61, 79, 91, 94, 98, 99, 100, 102
rings, 83
risk(s), 3, 10, 70, 72, 77, 105, 120
robust broadcast, 49, 50
roots, 14
royalty, 89, 90, 98, 118, 119, 120, 122
rules, 4, 6, 9, 10, 11, 14, 19, 34, 35, 38, 39, 40, 43, 48, 58, 59, 60, 63, 77, 78, 79, 123
rural areas, 71

S

safe haven, 14
Samsung, 30, 73
satellite service, 9
Satellite Television Extension and Localism Act of 2010 (STELA), 5
school, 116
science, 11, 135
scope, 20, 63, 135
self-interest, 139
Senate, 54, 55, 56, 114
servers, 136
service provider, 31, 61, 95
shape, 35, 41, 42, 76
shelter, 108
showing, 99, 117
signals, 5, 39, 40, 44, 53, 54, 55, 56, 60, 61, 62, 64, 79, 105
Silicon Valley, 29
skilled workers, 139

Slovakia, 103
small businesses, 69
socialization, 136
software, 15, 26, 76
solution, 41
specifications, 107
speculation, 26, 63
spending, 121, 134
spin, 89
Spring, 109
Sprint, 68
stakeholders, 118
stars, 26
state(s), 2, 3, 44, 69, 77, 106
statutory provisions, 53
storage, 68, 135
storms, 105
structural changes, 38
structure, 34, 118, 120
style, 112, 113
subscribers, 4, 5, 6, 9, 13, 14, 15, 16, 17, 18, 19, 20, 21, 23, 26, 36, 43, 53, 54, 55, 56, 67, 72, 74, 75, 78, 82, 85, 103, 106, 134
subscription networks, 53
subsidy, 101, 123
substitution, 40
supplier(s), 58, 59, 64
Supreme Court, 5
survival, 60
syndicated exclusivity rules, 4, 58

T

tactics, 13, 40
talent, 83
tanks, 137
target, 60
Task Force, 93
technological advances, 48
technological change, 78
technological progress, 137
technology(ies), 2, 4, 10, 11, 15, 17, 18, 20, 22, 26, 31, 34, 35, 39, 41, 42, 48, 51, 61, 67, 68, 74, 77, 79, 87, 88, 90, 95, 96, 100, 106, 112, 114, 121, 123, 127, 131, 134, 135, 137
teens, 112
telecommunications, 67
Telecommunications Act, 2
telephone, 63, 72, 73, 77, 78, 107, 134
telephone companies, 63, 78
telephones, 134
television stations, 47, 49, 57, 58, 59, 84
television viewing, 47, 57
tensions, 83
theatre, 74
theft, 95, 96, 100, 102
threats, 19
Title V, 36
tornadoes, 105
Toyota, 123
tracks, 88, 98
trade, 41, 69, 87, 93, 133
trade agreement, 41
trajectory, 101
transformation, 71, 102, 126, 130
transmission, 51, 52, 56, 82, 112, 113, 114, 123, 138
treatment, 22, 37, 56, 119
turp, 77

U

United Kingdom, 106
United States (US), 10, 15, 21, 25, 42, 44, 67, 79, 93, 99, 103, 107, 133
United, 10, 15, 21, 25, 42, 44, 67, 79, 93, 99, 103, 106, 107, 133
universe, 71, 134
universities, 139
urban, 72, 78

V

varieties, 61
Verizon, 2, 30, 38, 68, 73, 78, 107
vertical integration, 78
Vice President, v, vi, vii, viii, 1, 29, 65, 82, 125, 129

video market, vii, 2, 11, 27, 34, 35, 36, 38, 40, 41, 42, 49, 54, 56, 75, 76, 78
video programming, 6, 14, 15, 17, 20, 25, 35, 36, 37, 43, 44, 49, 54, 55, 72, 73, 75, 77, 78
videos, 83, 101
vision, 121

W

waiver, 85, 117
waking, 89
Washington, 26, 133, 134
watches, 30
water, 33
weakness, 59
wealth, 65, 76, 116
web, 9, 17, 23, 44, 73, 75, 77, 139
web service, 75
websites, 73, 95, 107, 114, 135, 139
Wi-Fi, 72, 76
wildfire, 116
wireless devices, 51, 85, 126, 127, 130, 131
wireless networks, 51, 68, 105
wireless technology, 52
Wisconsin, 30
witnesses, 1, 81
workers, 69, 139
World Wide Web, 15, 16
worldwide, 9, 98, 133

Y

young adults, 138
young people, 48